Mechanical Devices for the Electronics Experimenter

Other books by C.B. Rorabaugh

Circuit Design and Analysis: Featuring C Routines
Communications Formulas and Algorithms: For Systems Analysis and Design
Digital Filter Designer's Handbook: Featuring C Routines

Mechanical Devices for the Electronics Experimenter

Britt Rorabaugh

TAB Books
Division of McGraw-Hill
New York San Francisco Washington, D.C. Auckland Bogotá
Caracas Lisbon London Madrid Mexico City Milan
Montreal New Delhi San Juan Singapore
Sydney Tokyo Toronto

Published by TAB Books, a division of McGraw-Hill, Inc.

pbk 8 9 10 11 FGR/FGR 0 5 4 3 2 1
hc 6 7 8 9 10 11 12 FGR/FGR 9 0 0 9 8

Library of Congress Cataloging-in-Publication Data
Rorabaugh, Britt.
Mechanical devices for the electronics experimenter / by Britt Rorabaugh.
p. cm.
Includes index.
ISBN 0-07-053546-9 (h) ISBN 0-07-053547-7 (p)
1. Electronics—Amateurs' manuals. 2. Electronics—Experiments. 3. Mechanical movements. I. Title.
TK9965.R65 1995
629.8'92'078—dc20 94-36306
CIP

Acquisitions editor: Roland S. Phelps
Editorial team: Aaron G. Bittner, Book Editor
Robert E. Ostrander, Executive Editor
Jodi L. Tyler, Indexer
Production team: Katherine G. Brown, Director
Jan Fisher, Desktop Operator
Nancy K. Mickley, Proofreading
Toya Warner, Computer Artist
Brenda Wilhide, Computer Artist
Design team: Jaclyn J. Boone, Designer

0535477
EL1

To Joyce, Geoff, and Amber

Contents

Introduction

Projects that electronics experimenters undertake often have significant mechanical content in addition to their electronics content. In fact, for projects in the area of robotics, the electronics content is usually secondary to the mechanical content. There is a heavy emphasis on robotics applications, but this book is intended for **all** experimenters. Because it is intended for **experimenters**, this book contains ideas for things that can be tried out in a home shop. The analysis formulas and design rules provided are intended to give the reader a good foundation upon which to expand—primarily through experimentation and measurement. The formulas and design rules provided will **not** equip the reader to design a mechanism to a set of stringent specifications and then make precise analytic predictions as to how the finished mechanism will perform. These sorts of activities are the province of professional mechanical engineers—not amateur experimenters. A good strategy for using the information and ideas in this book consists of the following steps:

1. Design a little.
2. Build a little.
3. Test a little.
4. Repeat 1, 2, and 3 as needed.

Because the real fun comes from watching an original design evolve into a working reality, breaking the design, building, and testing into a series of small pieces reduces the wait between the fun parts.

1
CHAPTER

Basic mechanical principles

This chapter will explain the basic mechanical concepts that are used in the rest of the book. Some specific topics to be covered include: force, work, torque, power, friction, and simple machines.

Fundamental quantities

In almost every field of endeavor, there is a certain amount of tedious introductory material that must be forged into a foundation upon which the more interesting topics can be developed. For the study of mechanical devices, most of this introductory material is drawn from the field of physics. In turn, the structured study of physics is built upon a few fundamental concepts which include fundamental physical quantities like mass, length, and time; laws of motion; laws of force; and simple machines. This chapter is devoted to fundamental concepts of this sort.

Indefinables in mechanics

In mechanics, there are three fundamental indefinables that are not defined in terms of other quantities. Every other quantity of interest can be defined in terms of these other three. The units used to measure indefinable quantities are called fundamental units. These fundamental units were originally defined in terms of actual tangible physical things such as the length of a particular metal bar, the mass of a particular metal cylinder, etc. Units for other quantities of interest in mechanics are defined in terms of these fundamental quantities. Two of the indefinables are time and length. The third indefinable can be chosen as either mass or force. Because the choice between mass or force is somewhat arbitrary, we clearly need a way to define mass in terms of time, length, and force; or conversely, a way to define force in terms of time, length, and mass. The desired relationship is Newton's second law of motion:

$$\text{force} = \frac{\text{mass} \times \text{distance}}{\text{time}^2}$$

This relationship will be discussed in more depth later in this chapter.

Length

Length is a straightforward concept; anyone reading this book should already have a warm and fuzzy feeling about what length is. Today the internationally accepted fundamental unit of length is the meter. Any of us can go out and buy a meter stick if we wish to measure something in units of meters. Suppose, for the sake of argument, that two different meter sticks do not match exactly. Which one is correct? Strictly speaking, it is likely that neither one is *exactly* correct, but the one that is closer to correct is the one that most closely matches the international standard for the meter. Since the late eighteenth century, there has been a succession of platinum-iridium bars kept in a vault at Sèvres, France. Until 1960, the official international meter was defined as the distance between two fine scratches on one of these bars. In 1960, this standard was replaced by one based on the wavelength of light emitted by krypton-86. The precise definition was adopted as follows:

> The meter is the length equal to 1,650,763.73 wavelengths in vacuum of the radiation corresponding to the transition between the levels $2p^{10}$ and $5d^5$ of the krypton-86 atom.

The wavelengths of light are counted using a device called an interferometer. Within the metric system, other units of length based on the meter are obtained by multiplying or dividing by powers of 10.

Table 1-1 lists the prefixes that are used in the metric system to form multiples and submultiples of the basic units. For example, one thousandth of a meter (or 10^{-3} meters) is the same as one millimeter (1 mm). A thousand (or 10^3) meters is the same as one kilometer (1 km). The prefixes for multiples of the basic unit are derived from Greek roots, and the prefixes for submultiples of the basic unit are derived from Latin roots. The word "meter" itself comes from the Greek metron, meaning "measure."

In the United States, the yard is defined as 0.91440183 meter. In Britain, the yard is defined as 0.9144 meter, thus making it 1830 nanometers shorter than the US yard. To the home machinist, the difference is moot. Table 1-2 lists units of length that are likely to be of interest to an experimenter in electromechanics.

Time

Time is a familiar concept, but is perhaps difficult to define. We all have a feeling for how long a second, a minute, or an hour is; but we could debate forever the question, "What is time?" Let's steer clear of that discussion and just say that the international unit of time is the second. A second was originally defined as 1/86400 of a mean solar day. In 1960 this definition was changed to be 1/31,556,925.9747 of the tropical year 1900. Several years later, the second was redefined as:

> The duration of 9,192,631,770 periods of the radiation corresponding to the transition between two hyperfine levels of the cesium-133 atom.

Table 1-1. Metric prefixes

Prefix	Abbrev.	Multiplier
tetra-	T	10^{12}
giga-	G	10^{9}
mega-	M	10^{6}
kilo-	k	10^{3}
hecto-	h	10^{2}
deka-	da	10
deci-	d	10^{-1}
centi-	c	10^{-2}
milli-	m	10^{-3}
micro-	μ	10^{-6}
nano-	n	10^{-9}
pico-	p	10^{-12}
femto-	f	10^{-15}
atto-	a	10^{-18}

Table 1-2. Units of length

Unit	Abbrev.	Equivalent to
centimeter	cm	0.0328084 feet
		0.3937008 inch
foot	ft	0.3048 meter
		30.48 centimeter
		12 inches
inch	in	0.0254 meter
		2.54 centimeter
		25.4 millimeter
kilometer	km	0.6213712 US statute mile
meter	m	39.37008 inch
		3.28084 foot
		1.093613 yard
mil	—	0.001 inch
mile	mi	1609.344 meter
		1.609344 kilometer
		5280 feet
		1760 yard
yard	yd	

The metric prefixes for submultiples are commonly applied to the second to obtain millisecond (ms), microsecond (μs), nanosecond (ns), and picosecond (ps). The prefixes for multiple units could be (but rarely are) used for time measurements. It is more common to see a quantity such as 1275 seconds, rather than 1.275 ks.

Mass and weight

One of the first conceptual hurdles in any serious study of mechanics involves the distinction between mass and weight. Every object that we use in the construction of a mechanical gizmo has some weight that we can observe or measure by using a suitable scale. Intuitively, it seems as though weight should be an intrinsic property of matter; a one-pound chunk of metal should always weigh one pound, no matter what. However, we know this is not always the case. On the moon, things weigh less than they do on earth; in orbit, things are "weightless." Clearly, the weight of an object depends on the local effects of gravity. Unlike weight, mass is an intrinsic property of an object. Every object has some particular mass, and the weight that we observe is the force exerted on this mass by the Earth's gravity. If the object were taken to the moon, the weight would change, but the mass would remain the same. However the weights of objects are always proportional to their masses:

$$\text{weight} = k \times \text{mass}$$

where the constant of proportionality k, is the local gravitational constant. As we will discover shortly, k has units of acceleration ($\text{ft}\cdot\text{s}^{-2}$, $\text{m}\cdot\text{s}^{-2}$, etc.) and is often called the "acceleration due to gravity."

In physics, the difference between mass and weight is so important that different units are used for measuring the two. Forces, including weight, are expressed in *newtons*, *dynes*, *ounces*, or *pounds*; mass is expressed in *kilograms*, *grams*, or *slugs*. This distinction between mass and weight may seem contrived, and it is often difficult to accept at first. The average person doesn't know or care about the difference. As a consequence of metrification in the consumer marketplace, further confusion is introduced by packages of cake mix that read "net weight 21.5 ounces (609 grams)." We just said that ounces are used to measure force and grams are used to measure mass. Scientists would be happier if the package read "net weight 21.5 ounces (596,820 dyne)." The reasons for the apparent inconsistency are rooted in history. To a nonphysicist, the idea of weight seems more fundamental than the idea of mass. After all, daily commerce has for centuries been conducted on the basis of weight. In fact, when standardized systems of weights and measures were first established, they were based on weight rather than mass. The *British engineering* system of weights and measures was based on a unit of force (1 lb) and a unit of acceleration (1 $\text{ft}\cdot\text{s}^{-2}$). The unit of mass was then defined as the mass of a body whose acceleration is 1 $\text{ft}\cdot\text{s}^{-2}$ when a force of 1 lb is applied to the body. In 1799, under the auspices of the Academy of Sciences, a metal prototype kilogram was established as a standard of weight. In 1889, this original standard was replaced by a platinum-iridium cylinder. This particular alloy was chosen for its resistance to wear and tarnishing. This cylinder is still the world's standard kilogram, but it has been redesignated as a standard of mass rather than a standard of weight. This redesignation makes sense, because the weight of this cylinder or of any primary copies will vary at different places on the earth's surface due to local variations in gravity. However, the mass of the cylinder will remain constant.

Although physicists make a distinction between the concepts of weight and mass, the difference is not really that important to an amateur experimenter working on the surface of the earth. Therefore, in this book, we will assume that an object's mass is equal to its weight as measured on earth. Consistent with this approach, engineers commonly make use of a unit of 1 pound mass (lbm) that is defined as 0.45359237 kg or 0.031081 slug. Additional units for measuring mass are listed in Table 1-3. Conversely, a unit of 1 kilogram force (kgf) is defined as 9.80665 newton. Additional units for measuring force and weight are listed in Table 1-4.

Table 1-3. Units of mass

Unit	Abbrev.	Equivalent to
kilogram	kg	2.204 pound-mass
pound-mass	lbm	0.45359237 kilogram
		0.031081 slug
slug	—	32.174 pound-mass
gram	g or gm	10^{-3} kilogram
tonne	t	10^{3} kilogram
(metric ton)		2205 pound-mass
carat	c	0.2 gram

Table 1-4. Units of force or weight

Unit	Abbrev.	Equivalent to
dyne	dyn	2.248×10^{-6} pound-force
		10^{-5} newton
newton	N	10^{5} dyne
		0.102 kilogram-force
kilogram-force	kgf or kg_f	9.81 newton
kilopond	kp	(Note 1)
poundal	pdl	32.174 lbf
ounce (avoirdupois)	oz	0.0625 pound-force
		28.349527 gram
ounce (troy)	oz t	31.103481 gram
stone	st.	14 pound-force
ton (short)	ton (Note 2)	2000 pound avoirdupois
ton (long)	l.t.	2240 pound avoirdupois
sten or sthene	sn	1000 newton
dram	dr	3.887 gram force
grain	gr	0.064789918 gram force

Notes:

(1) Kilopond is an alternative name for kilogram-force.

(2) Short ton is abbreviated s.t. when needed to emphasize difference between long ton and short ton.

Why are we making such a big deal about units? The reason is that the units we choose for measurement will have an impact on how we use many of the design formulas presented in later chapters. For example, Newton's second law of motion is stated mathematically as:

$$F = ma \tag{1-1}$$

where F is force, m is mass, and a is acceleration. If these three quantities are expressed in consistent units, then Eq. 1-1 can be used exactly as is. One possible set of consistent units expresses force in pounds (lb), mass in slugs, and acceleration in (feet/second)/second or ft•s^{-2}. The force (in pounds) needed to accelerate a 2 slug mass at 5 f•s^{-2} is obtained by direct application of Eq. 1-1:

$$F = (2 \text{ slug})(5 \text{ m} \cdot \text{s}^{-2}) = 10 \text{ lb}$$

However, suppose we want to determine the force needed to accelerate a mass of 2 lbm at 5 ft•s^{-2}. The mass and acceleration are not in consistent units, so we need to convert the mass from lbm to slugs:

$$2 \text{ lbm} \times 0.031081 \frac{\text{slug}}{\text{lbm}} = 0.062162 \text{ slug}$$

Then we use this new value of mass in Eq. 1-1:

$$F = (0.062162 \text{ slug})(5 \text{ m} \cdot \text{s}^{-2}) = 0.3108 \text{ lb}$$

It may seem cumbersome at first, but writing formulas that expect quantities to be in consistent units is really the most general way to do things. The alternative would be sets of formulas, with a different specific formula for each combination of units that we might want to use. Table 1-5 shows just how complicated this alternate approach could become. Even though formulas in consistent units are generally preferred, for some engineering disciplines a specific formula is developed in situations where one particular set of inconsistent units is used repeatedly.

Table 1-5. Newton's second law expressed in formulas using mixed units

Force unit	Mass units	Accel. units	Formula
n	kg	m • s^{-2}	$F = ma$
dyn	g	cm • s^{-2}	$F = ma$
lb	slug	ft • s^{-2}	$F = ma$
lb	lbm	ft • s^{-2}	$F = \frac{ma}{32.174}$
kgf	kg	m • s^{-2}	$F = 9.80665\, ma$
lb	ozm	in • s^{-2}	$F = \frac{ma}{6177.408}$
oz	ozm	in • s^{-2}	$F = \frac{ma}{2.681}$
kgf	stone	ft • s^{-2}	$F = 18.9815\, ma$

Additional primary units

In addition to the units for the three indefinables of mechanics, there are several other units that are considered basic, but which can be partially defined in terms of the three indefinables.

Electric current

The unit of electric current is the *absolute ampere*, which occupies a strange position in that it can be defined in terms of the three indefinables of mechanics, but it is nevertheless defined as a primary unit in the SI system of measurements. The definition of an absolute ampere is:

> The ampere, unit of electric current, is the constant current which, if maintained in two parallel conductors of infinite length, of negligible cross section, and placed 1 meter apart in a vacuum, will produce between these conductors a force equal to 2×10^{-7} newton per meter of length.

This is a great conceptual definition, but a little tough to apply in practice; conductors of infinite length and negligible cross section are hard to come by. The standard is actually implemented using devices called current balances that can measure currents accurately to within a few parts per million. An earlier ampere standard, called the *international ampere*, was defined as the "unvarying current which, in 1 second, deposits 0.001118 gram of silver from an aqueous solution of silver nitrate [$AgNO_3$]." The absolute ampere is equal to 1.000165 international ampere.

Luminous intensity

The SI unit of luminous intensity is the *candela*. The laboratory procedure for implementing the standard involves a pool of molten platinum (at a temperature of 2045 K) at the bottom of a narrow well constructed of thorium dioxide. The light emitted from this well is then compared to the light from a precision electric lamp. The voltage supplied to the lamp is adjusted until the illuminations from the well and the lamp appear to be equal. The lamp is then deemed to be a standard 1-candela lamp.

Temperature

The SI unit of temperature is the *kelvin* (K). At the Thirteenth General Conference on Weights and Measures in 1967, the following definition of the kelvin was adopted: "The kelvin, a unit of thermodynamic temperature, is 1/273.16 of the triple point of water."

This definition is of limited practical use. In practice, precision temperature determination makes use of the International Practical Temperature Scale (IPTS). To the amateur experimenter, all that really matters are the grade-school definitions of kelvin, Celsius, and Fahrenheit temperatures. The important points on these temperature scales are summarized in Table 1-6.

Table 1-6. Comparison of Fahrenheit, Celcius, and Kelvin temperature scales

Event	°F	°C	K
Water boils	212	100	373.16
"Standard" temperature	68	20	293.16
Water freezes	32	0	273.16

Note:

A change of 1°**F** equals a change of $\frac{5}{9}$°**C**.

A change of 1°**C** equals a change of 1.8°**F**.

Vector arithmetic

Forces, acceleration, and velocity are vector quantities. This means that they have a direction as well as a magnitude. Consider the situation depicted in Fig. 1-1. A force of 10 pounds is being applied to the block in a direction that is 30 degrees from horizontal. Intuitively, we know that some of this force acts vertically, tending to push the block into the surface on which it rests. Also, some of the force will act horizontally, tending to push the block towards the right. Often, in the design and analysis of mechanical systems we would like to be able to calculate just how much of a force acts horizontally and how much of a force acts vertically. Vector arithmetic provides us with a mathematical tool for doing just that.

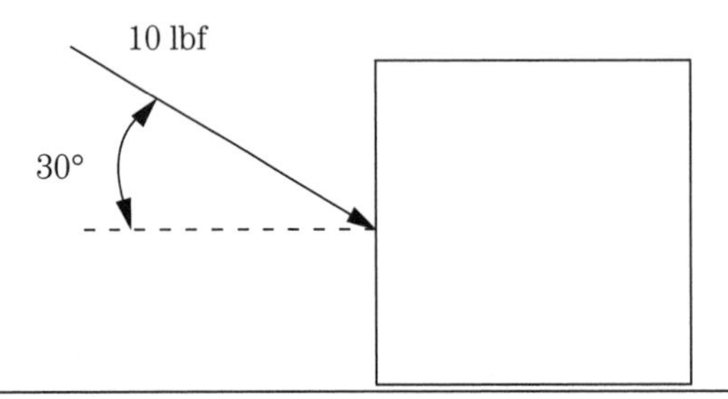

1-1 Illustration of force as a *vector* quantity. The force acting on the block has both a magnitude (10 lbf) and a direction (30° from horizontal).

Vectors

Any quantity that has both a magnitude and a direction is called a *vector quantity*, or simply a vector. A vector can be graphically represented by an arrow that points in the appropriate direction and that has a length which is proportional to the magnitude of the quantity being represented. Any arrow representing a vector can be placed in *standard position*, with its tail located at the origin of the cartesian plane as shown in Fig. 1-2.

It will be possible to locate a vertical line at some point along the horizontal axis such that the line passes through the tip of the vector. The portion of the axis from the origin to the point where the vertical line crosses the axis is called the vector's *horizontal component*. Similarly, it will be possible to locate a horizontal line at some point along the vertical axis such that the line passes through the tip of the vector. The portion of the vertical axis from the origin to the point where the horizontal line crosses the axis is called the vector's *vertical component*. To emphasize the fact that "magnitude-only" quantities are not vectors, such quantities are usually referred to as scalar quantities within the context of vector arithmetic. In printed material, vectors are distinguished by the use of boldface: a, b, c are scalars; **a**, **b**, **c** are vectors. In handwritten material, vectors are distinguished by using either an over-bar or over-arrow: $\bar{a}$, $\bar{b}$, $\bar{c}$; $\vec{a}$, $\vec{b}$, $\vec{c}$; or $\overleftarrow{a}$, $\overleftarrow{b}$, $\overleftarrow{c}$. The magnitude of a vector **a** is denoted as $|\mathbf{a}|$ and the angle is noted as arg(**a**).

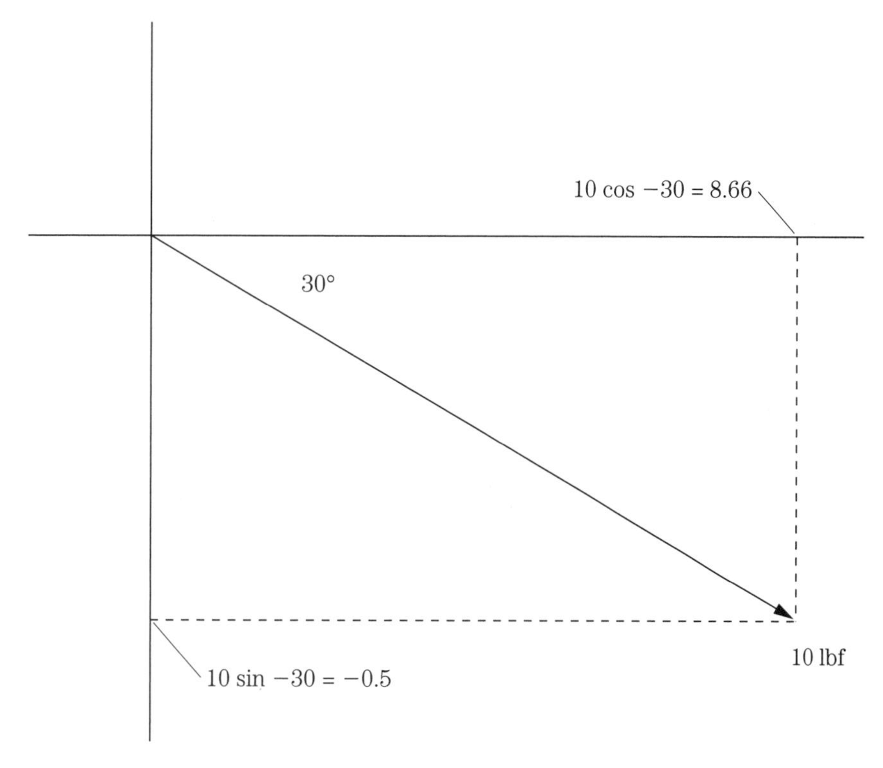

1-2 Force from Fig. 1-1 depicted as a vector in standard position in the Cartesian plane.

Resolving a vector into horizontal and vertical components

The projections of a vector onto the horizontal and vertical axes can be obtained graphically, as was indicated in Fig. 1-2, or these components can be calculated using the following formulas:

$$a_x = |\mathbf{a}| \cos (\arg(\mathbf{a}))$$

$$a_y = |\mathbf{a}| \sin (\arg(\mathbf{a}))$$

If a vector is described in terms of its x and y components, the vector's magnitude and phase can be obtained using

$$|\mathbf{a}| = \sqrt{a_x^2 + a_y^2}$$

$$\arg (\mathbf{a}) = \tan^{-1}\left(\frac{a_y}{a_x}\right)$$

Vector addition

The addition of two vectors **a** and **b** is accomplished as follows:

1. Place vector **a** in standard position and determine its vertical and horizontal components.
2. Place vector **b** in standard position and determine its vertical and horizontal components.
3. Add the vertical component of **a** to the vertical component of **b**. The result is the vertical component of **a+b**.
4. Add the horizontal component of **a** to the horizontal component of **b**. The result is the horizontal component of **a+b**.
5. Calculate the magnitude and angle of the result using the horizontal and vertical component results obtained in steps 3 and 4.

Motion

Motion will play a part in almost any mechanism of interest to the experimenter. The study and analysis of motion is one of the constituents of modern mechanics. The study of motion usually begins with motion along a line, proceeds to motion in a plane, and finally ends up with motion through 3-dimensional space.

Speed and velocity

Speed is simply the rate at which something traverses a distance or length. Speed is expressed in units of length per unit time; miles per hour, feet per second, furlongs per fortnight, etc. For a car to travel at a speed of 60 mph it doesn't have to travel 60 miles or travel for an hour. The car simply has to travel at a rate such that it would cover 60 miles if it traveled at the same rate for one full hour.

Usually, in scientific and engineering contexts, the term velocity is used instead of speed in situations where a direction of movement as well as a rate of movement is being discussed. In other words, velocity is a vector quantity. All the usual rules of vector arithmetic apply to velocities.

Example 1

A model rocket is moving in a straight line that is 30° from the vertical, as shown in Fig. 1-3. The rocket's ground track is moving from point A to point B at a rate of 50 ft/sec. What is the rocket's speed measured along the path AC?

Solution: We can draw a vector diagram as shown in Fig. 1-4. We know that a vector B has a magnitude of 50 ft/sec, and using the rules of trigonometry we can determine that:

$$\cos 60° = \frac{|B|}{|C|}$$

or

$$|C| = \frac{|B|}{\cos 60°}$$

$$= \frac{50}{0.5} = 100 \text{ ft/sec}$$

Rotational speed

In many situations we will be concerned with the speed of something such as a motor, wheel, or gear that is rotating. Rotational motion is usually expressed in terms of either revolutions per unit time [revolutions per minute (rpm), revolutions per second (rps)] or in terms of angular distance traveled per unit time (degrees per second, radians per minute, etc.)

Acceleration

Acceleration is the rate at which speed changes. Acceleration is expressed in units of speed per unit time or units of distance per time squared. For example, if at some instant a robot is moving forward at a rate of 8 ft/sec, and then one second later it is moving forward at a rate of 10 ft/sec, it has undergone an acceleration of 2 feet per second per second. There are several different ways of indicating this notationally: 2 ft/sec/sec, 2 ft/sec^2, or 2 $\text{ft}\cdot\text{sec}^{-2}$. The last of these three is now the "preferred" notation, but the other two forms can still be found in many older references.

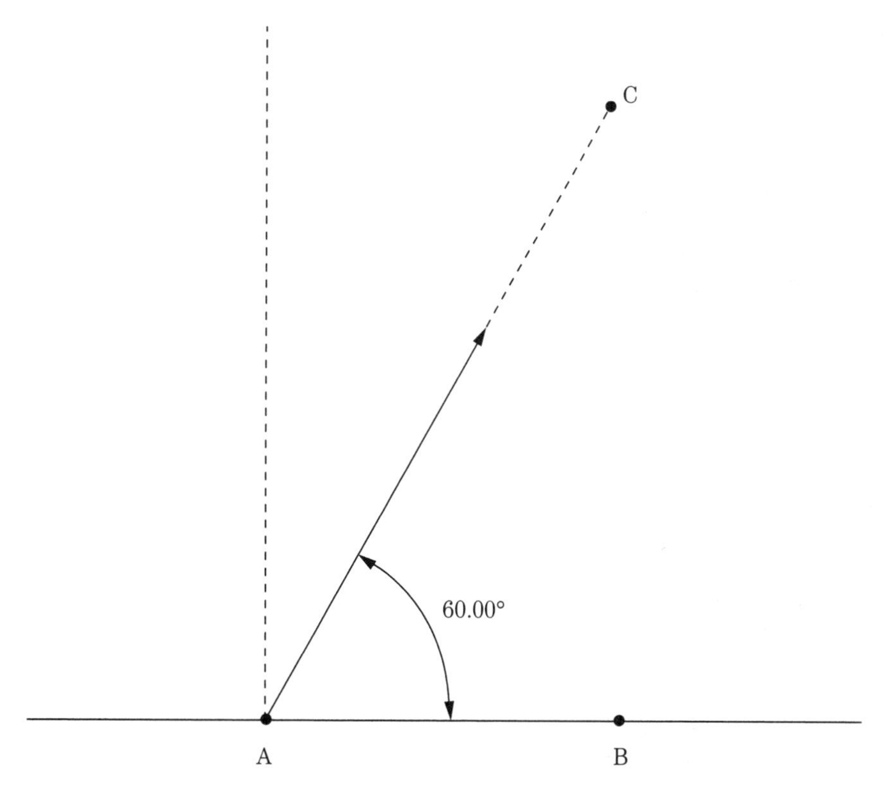

1-3 Model rocket trajectory for Example 1.

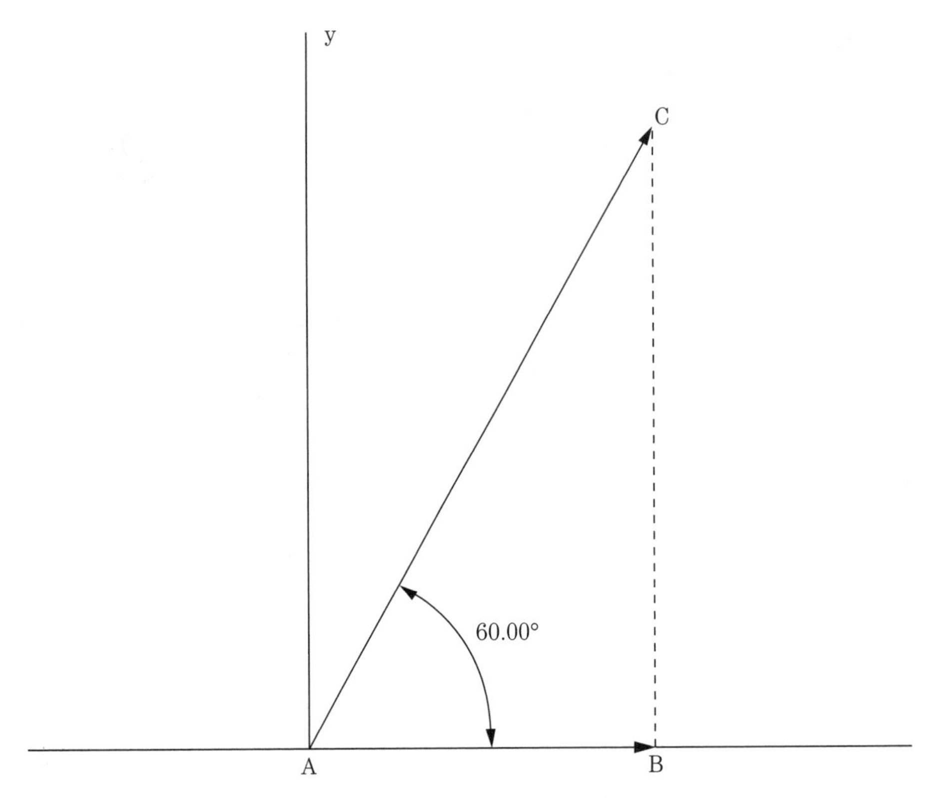

1-4 Vector diagram for Example 1.

Motion along a line

Motion along a line, or one-dimensional motion, is a good place to explore the relationship among displacement, speed, velocity, and acceleration. For a mathematical treatment of motion, we can arbitrarily assign coordinates to positions along a line, which as shown in Fig. 1-5, will make the line look like the "number line" used in elementary school arithmetic classes. The units of distance can be feet, inches, meters, miles, or whatever else makes sense in the context of the situation being analyzed. The displacement of an object is simply the difference between its ending coordinate and starting coordinate.

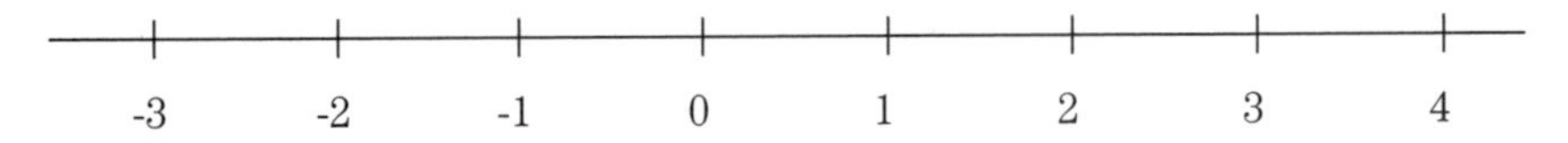

1-5 Coordinate axis for one-dimensional motion.

Example 2

Consider a nonsteerable robot that is capable of only straight-line motion in a forward or backward direction. As already stated, the assignment of coordinates can be arbitrary, but it is a common practice to assign coordinate zero to the object's location at time zero. Following this practice, and choosing to indicate positions in meters relative to the zero position, we can make the sketch shown in Fig. 1-6.

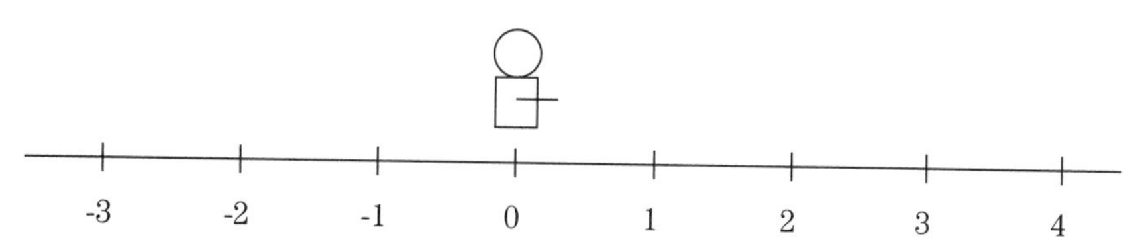

1-6 Robot's position at time zero.

If, at time t = 1 sec, the robot is located at the 2 meter position, we can say that there has been a displacement of 2 – 0 = 2 meters to the right. If at time 2, the robot is located at the –3 meter position, we can say that from time 1 to time 2 there has been a displacement of 5 meters to the left. Usually, when using this sort of coordinate line, displacements to the right are denoted as positive numbers and displacements to the left are denoted as negative numbers. Thus from time 0 to time 1, there has been a displacement of (–3) – 2 = –5 meters. We can also say that from time 0 to time 2, the overall net displacement is (–3) – 0 = –3. Besides going from position to displacement, we can go from displacement to position. At time 2, the robot is at the –3 meter position. If from time 2 to time 3 the robot undergoes a displacement of –4 meters, what will its position be at time 3? The answer is simply:

$$(-3) + (-4) = -7$$

The relationships developed in this example can be summarized as:

$$\text{displacement} = \text{end position} - \text{start position}$$
$$\text{end position} = \text{start position} + \text{displacement}$$
$$\text{start position} = \text{end position} - \text{displacement}$$

This coordinate system may seem like a lot of trouble at this point, but its worth will become more apparent as we consider velocity and acceleration.

The various robot motions studied in example 2 can be summarized using a plot position versus time, as shown in Fig. 1-7. In the figure, dots have been used to indicate the robot's position at times 0, 1, 2, and 3. We might be tempted to connect these dots with straight line segments, but this would say something; probably the wrong something; about how the robot moves from position to position.

A straight line between two position points would indicate that the robot moved between the two points at a constant speed, with this speed being the absolute value of the displacement divided by the time interval between the positions. Specifically, between times 1 and 2, the speed would be:

$$\frac{|-3-2|}{2-1} = \frac{5}{1} = 5 \text{ m/sec}$$

(Assuming of course that our times are actually in seconds, and our distances in meters.) So far, we haven't really said anything about the time units. They could be in seconds, milliseconds, minutes, hours, or any other convenient units. The speed between times 0 and 1 is given by:

$$\frac{|2-0|}{1-0} = \frac{2}{1} = 2 \text{ meters/sec}$$

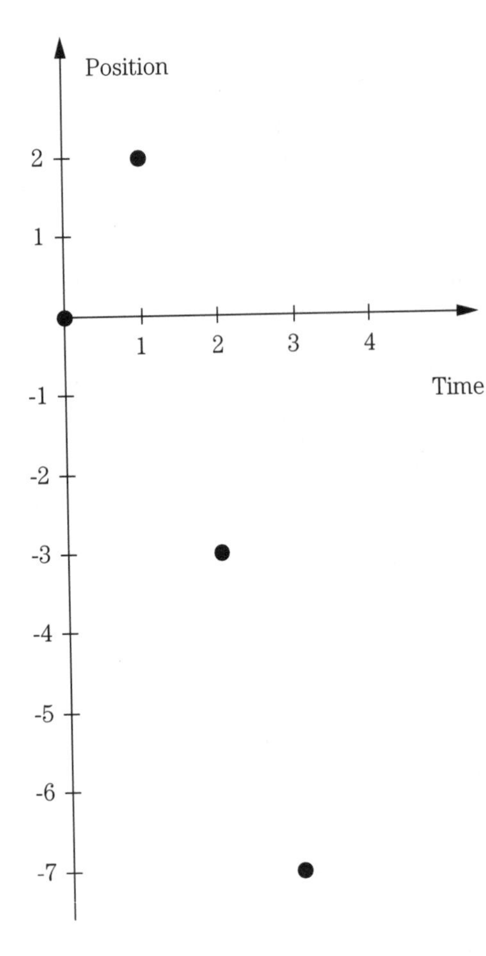

1-7
Plot of robot position versus time.

Just as we used a minus sign to denote displacement to the left, we can use a minus sign to denote speeds associated with displacements to the left. When speed has a direction associated with it, it is usually called *velocity*.

The robot motion depicted by Fig. 1-8 would require the robot's velocity at time 1 to instantly change from 2 to –5. Nothing in the real world can instantaneously change its velocity this way. A more realistic motion pattern is shown in Fig. 1-9. The curves connecting the position dots indicate that the robot's velocity is not constant. As drawn, this figure indicates that the velocity at each integer-valued time is zero.

A blown-up view of times 0 and 1 is shown in Fig. 1-10. The curved portion immediately to the right of each position dot indicates that the robot is accelerating from a velocity of zero, and the curved part immediately to the left of each position dot indicates that the robot is decelerating back to a velocity of zero.

The nearly straight path connecting each pair of acceleration and deceleration curves indicates the portion of time over which the robot's velocity remains nearly constant. The robot's velocity along this path of constant-velocity travel is simply the distance traveled divided by the time duration of this path, or in other words, the velocity is equal to the slope of the straightline portion of the curve. During the period

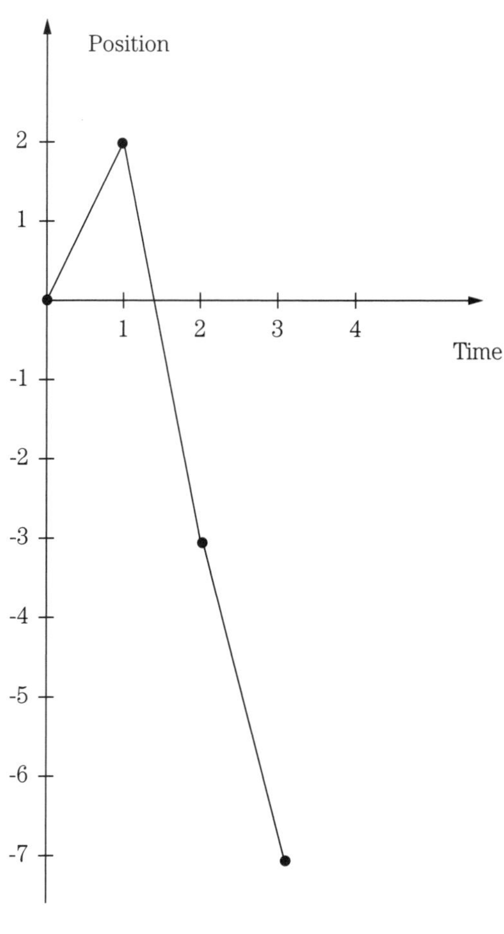

1-8
Plot of robot position versus time assuming piecewise constant velocities.

of acceleration, the robot's velocity is smoothly changing from a value of zero to this constant value. The *instantaneous* velocity is defined at every point along this curve, but it is difficult to figure out the value for any specific point without the use of differential calculus.

Mathematical approach

For those readers familiar with calculus, the relationships among position, velocity, and acceleration can be stated quite succinctly. If an object's position is known as a function of time, say $x(t)$, the object's velocity is the derivative of position, and acceleration is the derivative of velocity:

$$\text{position: } x(t)$$

$$\text{velocity: } v(t) = \frac{\mathrm{d}}{\mathrm{dt}}\, x(t)$$

$$\text{acceleration: } a(t) = \frac{\mathrm{d}}{\mathrm{dt}}\, v(t) = \frac{\mathrm{d}^2}{\mathrm{dt}^2}\, x(t)$$

For those readers not familiar with calculus, we present a few "calculus-free" results for some of the more important cases. As we observed previously, when an ob-

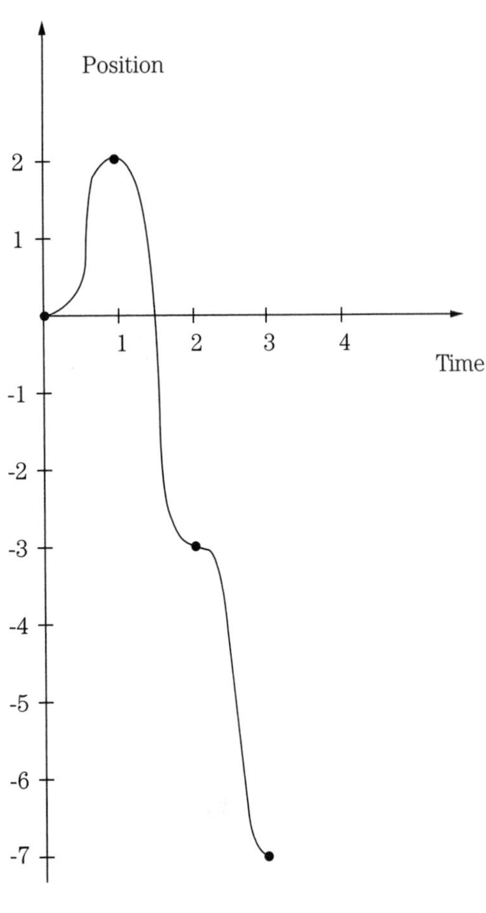

1-9
Plot of robot position versus time assuming gradual changes in velocity.

ject's position versus time can be represented by a straight line, the object is travelling at a constant velocity, with the velocity equal to the slope of the line. Because the velocity remains constant, the acceleration is obviously zero.

Mechanical quantities

Force, work, power, and energy are familiar words often used by nontechnical folks in everyday conversation. However, to the mechanical designer, these words each have a precise meaning, and this meaning is sometimes at odds with the meaning attached by common everyday usage. This section defines and explains the various quantities of interest to mechanical designers.

Force

Force can be thought of as a tendency to cause motion; or more precisely, a tendency to cause acceleration. For example, the force of gravity tends to accelerate objects toward the center of the earth; in fact, it will accelerate them unless the objects are restrained in some way. The amount of force required to cause an unre-

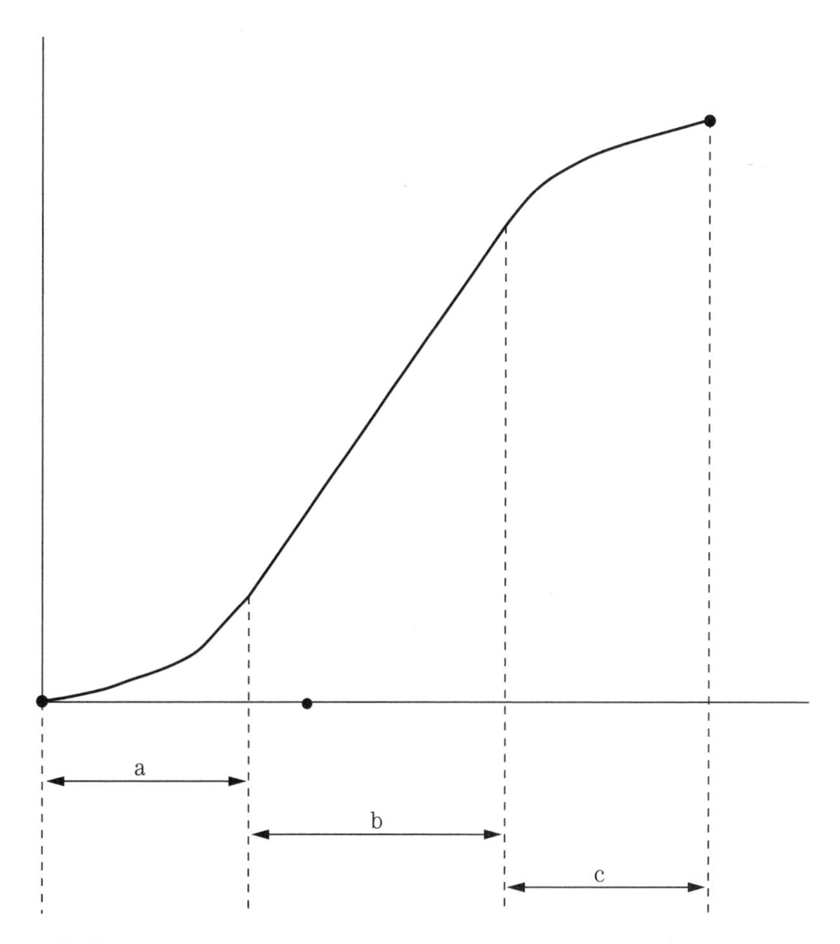

1-10 Enlarged detail of position plot between times 0 and 1. During (a) robot is accelerating, during (b) robot is traveling at constant speed, during (c) robot is decelerating.

strained object to accelerate is equal to the product of the object's mass and the amount of acceleration. Remember Newton's second law of motion?

$$F = ma$$

For example, if mass is in kg and acceleration is in $m \cdot s^{-2}$, then force will be in newtons. If mass is in slugs and acceleration is in $ft \cdot s^{-2}$, then force will be in pounds. If mass and acceleration are not in consistent units, the appropriate conversions must be performed prior to using Eq. 1-1. Downward acceleration due to earth's gravity is conventionally denoted as g. The value of g varies both with position on the earth and with altitude above the earth. Nevertheless, there is an arbitrary "standard" for gravitational acceleration that has been adopted by the International Committee on Weights and Measures. This value is 9.80665 $m \cdot s^{-2}$, or 32.174 $ft \cdot s^{-2}$. For the special case of gravity acting on a mass, we can modify the specific form of Eq. 1-1 to the specific form:

$$w = gm$$

where

w = weight

g = acceleration due to gravity

m = mass

The sets of consistent units for force, mass, and acceleration in the mks, cgs, and engineering systems are listed in Table 1-7. Conversions between various units of force are listed earlier in this chapter, in Table 1-2.

Table 1-7. Consistent units for Newton's second law

System	Force	Mass	Acceleration
mks (SI)	newton (N)	kilogram (kg)	$m \cdot s^{-2}$
cgs	dyne (dyn)	gram (g)	$cm \cdot s^{-2}$
engineering	pound (lb)	slug	$ft \cdot s^{-2}$

In addition to a magnitude, a force will also have a direction. Quantities that have a magnitude and direction are called vector quantities, and are usually depicted by arrows, as in Fig. 1-11. The length of the arrow used to depict a vector is proportional to the magnitude of the vector quantity. Forces can be added using the rules of vector addition, even if the forces are not acting in the same direction.

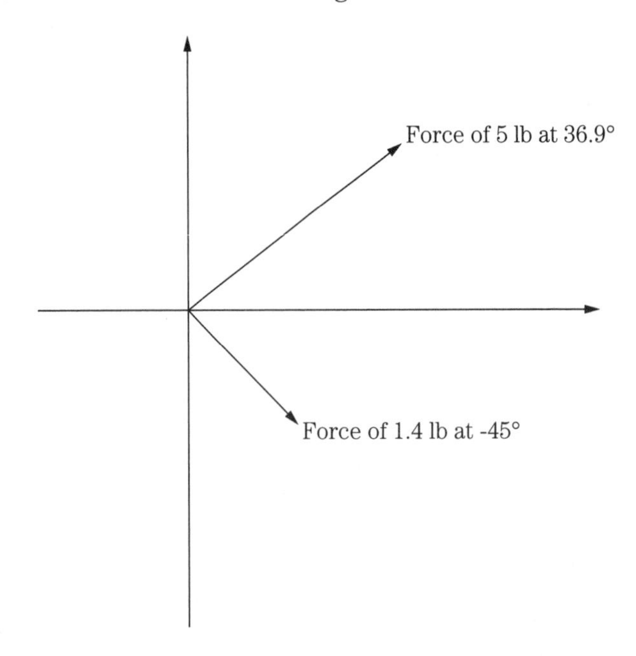

1-11 Vectors in the cartesian plane.

Example 3

Assume that a constant force of 1.5 newtons is applied to a wheeled cart on a level surface as shown in Fig. 1-12. Neglecting friction, what will the acceleration be if the cart has a mass of 7 kg?

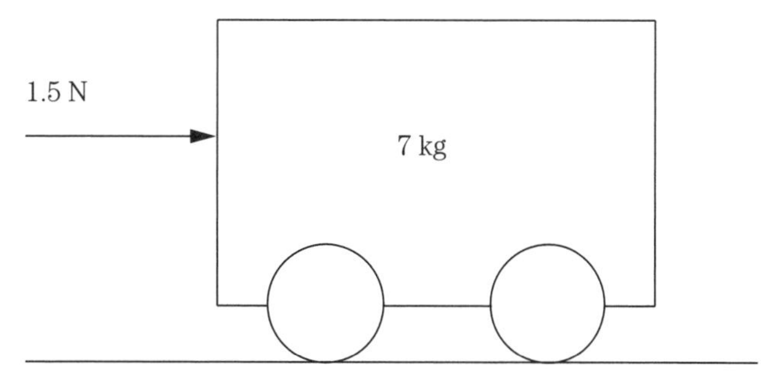

1-12 Cart for Examples 2 and 3.

Solution: Using Eq. 1-1, we have

$$a = \frac{F}{m} = \frac{1.5\ \text{N}}{7\ \text{kg}} = 0.21429\ \text{m}\cdot\text{s}^{-2}$$

Example 4

Assume that a constant force is applied as shown in Fig. 1-12 to a cart that has a mass of 20 lbm. How large must the force be to cause the cart to accelerate 3 ft•s^{-2}?

Solution: The consistent engineering unit of mass for use in Eq. 1-1 is the slug, which is equal to 32.174 lbm. Converting 20 lbm to slugs we obtain

$$20\ \text{lbm} \times \frac{1\ \text{slug}}{32.174\ \text{lbm}} = 0.62162\ \text{slug}$$

Therefore,

$$\begin{aligned} F &= (0.62162)(3) \\ &= 1.86486\ \text{lbf} \end{aligned}$$

How forces are measured

Although a balance can be used to measure the force exerted by gravity on a mass, they are generally not very useful for measuring forces. The traditional force-measuring tool is a spring balance. As shown in Fig. 1-13, a spring balance consists of a coil spring in a suitable housing with a pointer attached to one end of the spring. This pointer moves over a scale on the housing as forces applied to the end of the spring cause it to increase in length. Depending on the "elasticity" of the spring, the scale can be calibrated in grams, ounces, or pounds. It is surprisingly difficult to find a source of ready-made spring balances. They are still a staple in high school physics labs, and companies that sell scientific equipment to schools stock spring balances in several different force ranges. It is also possible to make spring balances in the home shop.

Equilibrium

Consider the rigid object shown in Fig. 1-14. Two forces, F_1 and F_2, are applied from exactly opposite directions. If the magnitude of F_1 and F_2 are exactly equal, the forces will "cancel" each other and the object will behave as if no forces at all were being applied. However, if say F_1 is greater than F_2, then the object will behave as

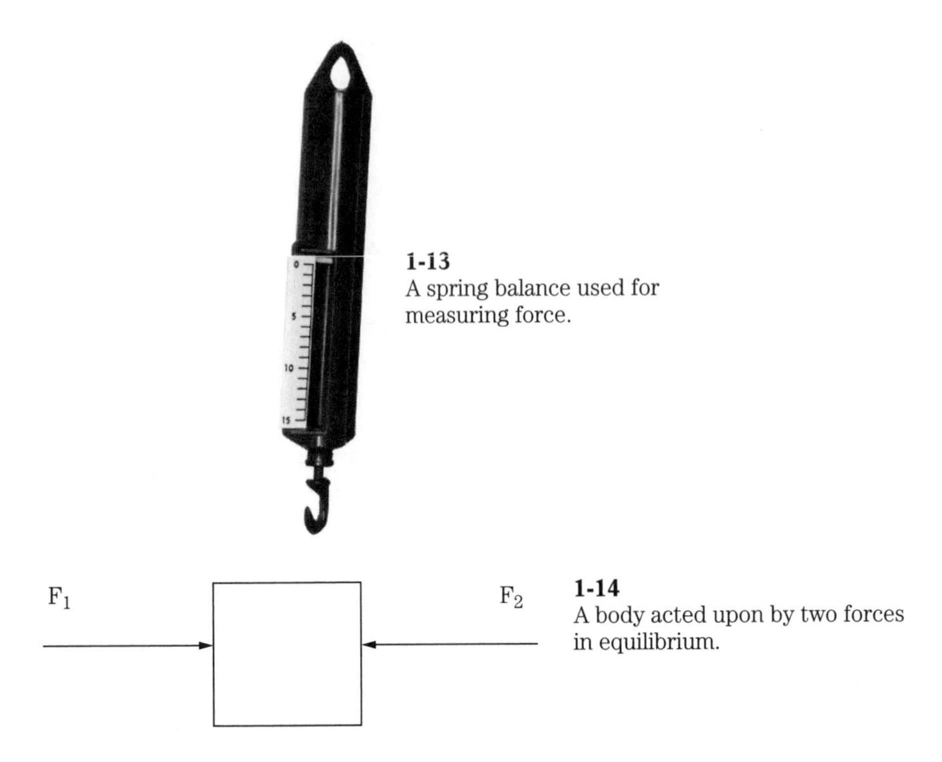

1-13
A spring balance used for measuring force.

1-14
A body acted upon by two forces in equilibrium.

though a force of magnitude $|F_1 - F_2|$ is applied in the direction of F_1. Conversely, if F_2 is greater than F_1, then the object will behave as though a force of magnitude $|F_1 - F_2|$ is applied in the direction of F_2. In short, when multiple forces are acting on an object, the object will behave as if it were being subjected to a single force equal to the vector resultant of all the separate forces.

Example 5

Consider the object suspended as shown in Fig. 1-15. Gravity pulls down on the object with a force of 50 lbf. Suspension cable **A** exerts a force of 100 lbf at an angle of 120° as shown. For the object to be in equilibrium, what force must be exerted by cable **B**?

Solution: We can resolve the force exerted by cable **A** into vertical and horizontal components as depicted in Fig. 1-16. For the body to be in equilibrium, the upward force F_{Ay} exerted by cable **A** must be equal in magnitude and opposite in direction to the downward force exerted by gravity. Using the conversions given earlier in this chapter, we can write:

$$
\begin{aligned}
F_{Ay} &= F_A \sin 30° \\
&= (100)(0.5) \\
&= 50 \text{ lbf}
\end{aligned}
$$

This result satisfies the requirement that $|F_{Ay}| = |F_g|$. Again using the conversions given earlier, we can write:

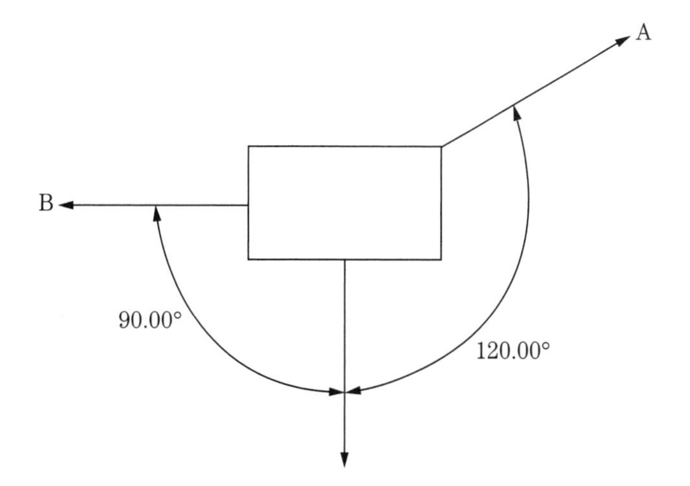

1-15 A body acted upon by three forces in equilibrium.

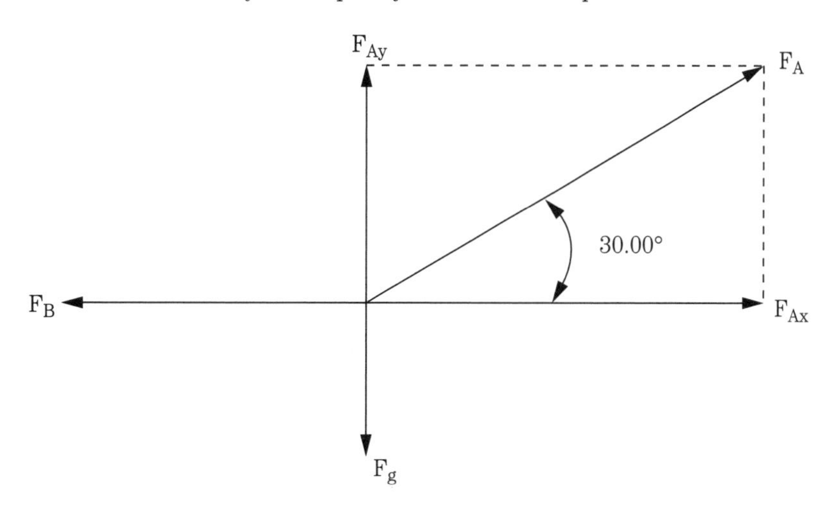

1-16 Vector diagram for resolving force **A** into vertical and horizontal components.

$$\begin{aligned} F_{Ax} &= F_A \cos 30° \\ &= (100)(0.866) \\ &= 86.6 \text{ lbf} \end{aligned}$$

Therefore, F_B must also be equal to 86.6 lbf in a leftward direction to balance the rightward force F_{Ax}.

Work and energy

Work can be thought of as force applied through a distance. A force F that moves something through a distance s performs work W given by the product of F and s:

$$W = Fs \tag{1-2}$$

The units of work are often expressed as combinations of other units; newton•meter, dyne•centimeter, foot•pound, etc; but there are also single units of work, as shown in Table 1-8. Energy is measured in the same units as work. In fact, work and energy can be viewed as just different forms of the same stuff. Electromechanical devices like motors, generators, and solenoids are designed to convert one form of energy into another. A motor converts electrical energy into mechanical work. A generator converts mechanical work into electrical energy. For a motor to perform x foot•pounds of mechanical work, we need to supply at least x foot•pounds of electrical energy as input to the motor. We say "at least" because all real world devices operate at something less than 100 percent efficiency. Therefore, some of the input energy is wasted (usually in the form of heat) and is not available for conversion into mechanical work. So really, we should say that a motor converts electrical energy into mechanical work and heat. Heat is just another form of energy and is measured in the same units as work or energy.

Table 1-8. Units of work

Unit	Abbrev.	Equivalent to
foot-pounds	ft • lb	5.05×10^{-7} horsepower-hours
		0.1383 kilogram-meters
		3.766×10^{-7} kilowatt-hours
Btu	Btu	778.3 foot-pounds
		1054.8 joules
ergs	ergs	10^{-7} joules
joules	J	0.7376 foot-pounds
kilowatt-hours	kWh	3413 Btu
		2.655×10^{6} foot-pounds
		3.6×10^{6} joules
ounce-inches	oz • in	7.062×10^{-3} newton-meters
newton-meters	N • m	1 joule
		0.7376 foot-pounds

Power

Power is the rate of doing work, or the rate of producing or using energy. The standard SI unit of power is the *watt* (W), which is equivalent to 1 *joule per second*. In terms of electrical power, 1 W is also equal to 1 volt•ampere (VA). When working in cgs units, power is not given its own unit; instead, it is expressed in ergs per second. A more traditional unit of mechanical power is the horsepower (hp), which equals 550 ft•lb/s. Conversions between various units of power are listed in Table 1-9.

Torque

How do we use the concept of force to characterize the turning effort produced by a motor? Consider the motor shown in Fig. 1-17.

Table 1-9. Units of power

Unit	Abbrev.	Equivalent to
horsepower	hp	42.41 Btu per minute
		3.3×10^4 foot-pounds per minute
		0.745 kilowatts
watts	W	44.26 foot-pounds per minute
		1.341×10^{-3} horsepower
foot-pounds per minute	ft • lb • min^{-1}	3.030×10^{-5} horsepower
		2.26×10^{-2} watts

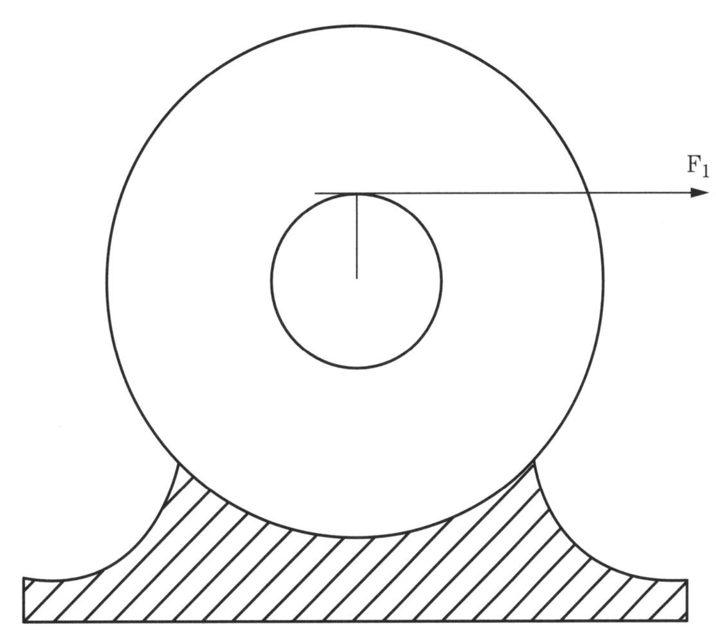

1-17 Turning force of a motor can be measured tangentially to the shaft.

We could measure the tangential force produced by the motor at some point on the circumference of the shaft as shown in the figure. Now let's cut down the shaft on a lathe to obtain the configuration shown in Fig. 1-18.

Or, alternatively, we could attach a pulley to the shaft as shown in Fig. 1-19 and measure the tangential force produced at the circumference of the pulley. When the three different motor configurations are run under identical conditions, the forces F_1, F_2, and F_3 will all be different.

Specifically, we can say that $F_3 < F_1 < F_2$. For each of these cases, the radial distance from the center of the shaft to the point of tangency is called the crank radius. How do we decide on the "correct" crank radius at which to measure the tangential force produced by the motor? It turns out that if we define a new parameter called

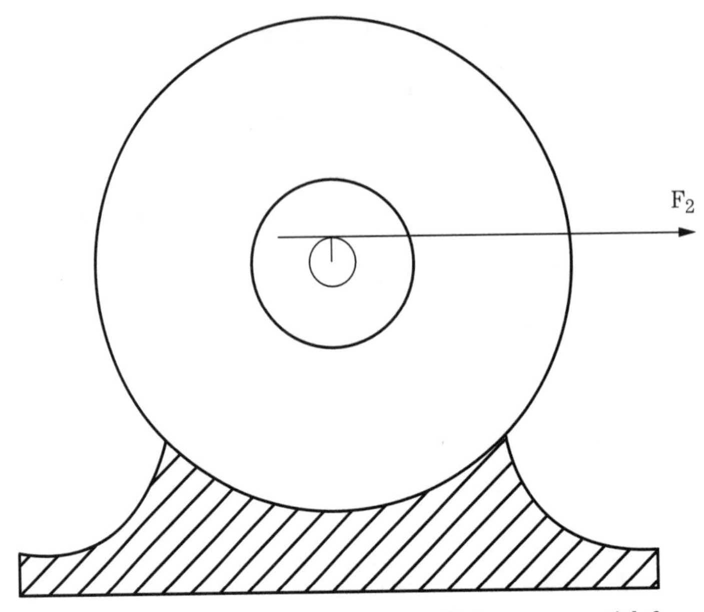

1-18 Motor with turned-down shaft could have tangential force measured at reduced-radius portion of shaft.

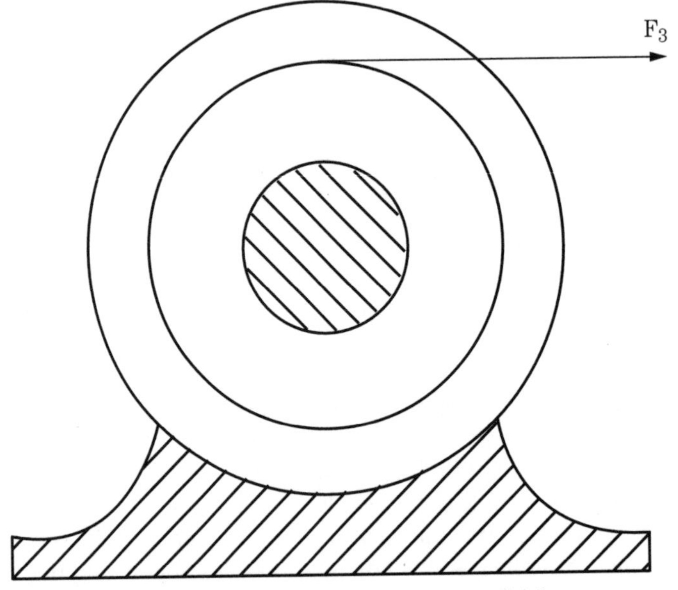

1-19 Motor with added pulley could have tangential force measured at increased area of pulley.

torque, that equals crank radius times tangential force ($\tau = rF$), we don't have to choose any particular radius because the torque will be constant, regardless of the radius at which it is measured. Because torque is the product of a force and a distance, it will have units that are the same as units of work or energy. This fact notwithstanding, torque is not work! Force times distance is work only when the force and distance are measured in the same direction. In our definition of torque, the force and distance are measured in directions that are orthogonal.

Torque, work, and power

Although torque and work are not the same, it is possible to establish a relationship between the two. A tangential force at a crank radius of r will move through a distance of $2\pi \cdot r$ each time the shaft makes one complete revolution, thus performing an amount of work given by

$$W = F \cdot 2\pi \cdot r$$

But the torque τ is equal to rF, so we can write

$$W = 2\pi \cdot \tau$$

This work per revolution is easily converted to a power figure, using the motor's speed in revolutions per second (f).

$$P = 2\pi \cdot \tau \cdot f \tag{1-3}$$

Example 6

Calculate the power (in hp) supplied by a motor running at 3000 rpm while delivering a torque of 700 foot•lb.

Solution: The desired information can be computed directly from Eq. 1-3, taking care to keep the units straight:

$$P = 700 \text{ ft}\cdot\text{lb} \times \frac{3000 \text{ rev}}{\text{min}} \times \frac{1 \text{ min}}{60 \text{ sec}} \times \frac{2\pi}{\text{rev}} \times \frac{\text{hp}\cdot\text{sec}}{550 \text{ ft}\cdot\text{lb}}$$

$$= 399.84 \text{ hp}$$

Friction

Consider the situation depicted in Fig. 1-20. A block having mass m is sitting on a horizontal surface. A horizontal force F_a is applied to the left side of the block. According to Newton's second law, this force should cause the block to accelerate toward the right at a rate of:

$$a = \frac{F_a}{m}$$

However, the block's tendency to accelerate rightward is impeded by friction between the two surfaces. Analytically it is convenient to treat friction as a force F_f that opposes the applied force F_a. As long as the magnitude of the applied force is

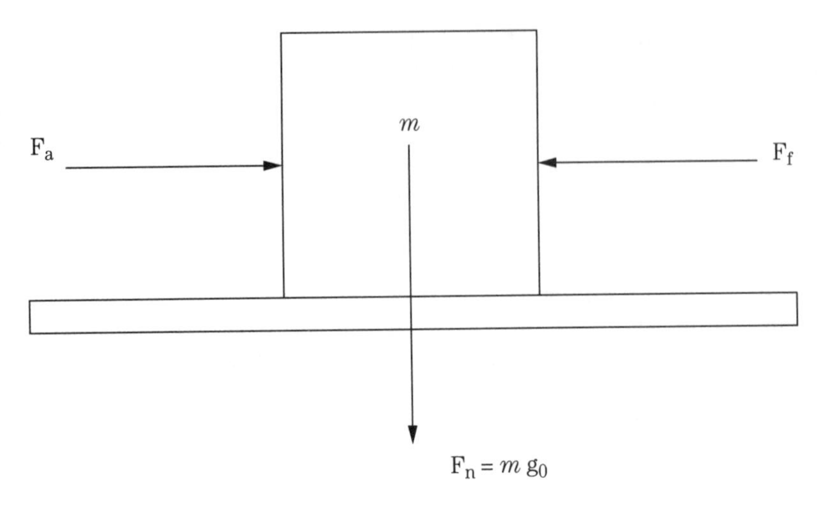

1-20 Illustration of forces for analyzing static friction.

less than some level F_{max}, the frictional force is considered to be equal in magnitude and opposite in direction to the applied force F_a. Under these conditions, F_f just balances F_a and no motion results. The level F_{max} depends on the characteristics of the surfaces and on the force with which the two surfaces are being pressed together. This "pressing-together" force is called the *normal force* F_n, because it is at right angles (normal) to the surfaces of interest. The effects of the surface characteristics are lumped into a single numeric value called the *coefficient of static friction* and denoted as μ_s. The value of F_{max} is given by:

$$F_{max} = \mu_s F_n$$

When the magnitude of the applied force F_a exceeds F_{max}, the block will begin to slide. Once the blocks are sliding, the friction force opposing the applied force depends on a different (and usually smaller) coefficient of dynamic friction μ_d. In order to keep the block moving, the applied force must exceed $\mu_d F_n$. In the usual cases where $\mu_d < \mu_s$, this means that there will be a bump in the plot of F_f versus F_a as shown in Fig. 1-21.

Levers

A lever is one of the so-called simple machines. A first class lever with its various parts labeled is shown in Fig. 1-22. A downward force on the effort arm will cause an upward force to be applied to the load.

This is one of the useful properties of levers: levers can be used to change the direction of an applied force. The *ideal mechanical advantage* (IMA) of a lever is defined as the ratio of the effort-arm length to the load-arm length:

$$\text{IMA} = \frac{r_E}{r_L}$$

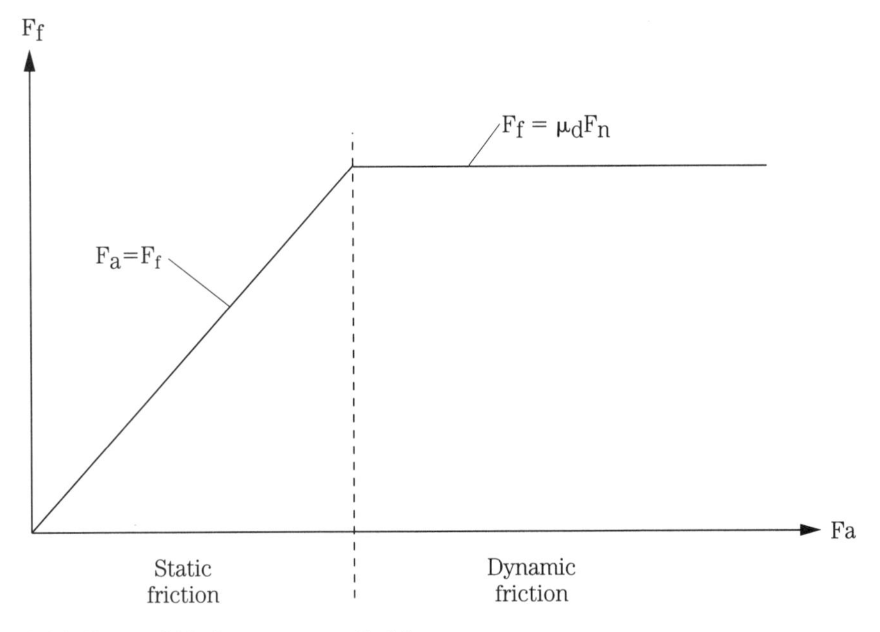

1-21 Force of friction versus applied force.

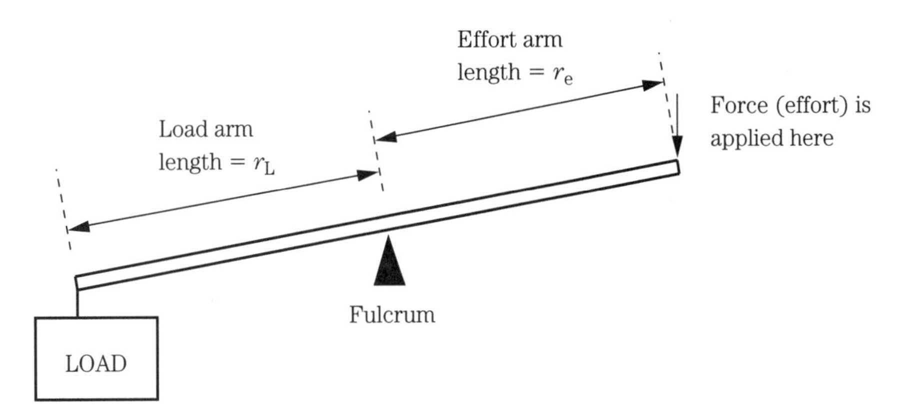

1-22 Parts of a first class lever.

The *actual mechanical advantage* (AMA) of a lever is defined as the ratio between force applied to the load and force applied to the effort arm:

$$\text{AMA} = \frac{F_L}{F_E}$$

The efficiency η of the lever is the ratio of AMA to IMA:

$$\eta = \frac{\text{AMA}}{\text{IMA}}$$

Friction at the fulcrum will cause the AMA to be less than the IMA, resulting in an efficiency less than 1. Whenever the fulcrum is closer to the load than it is to the effort, the IMA will be greater than 1, and (assuming $\eta = 1$) the force applied to the load will be greater than the applied effort. This illustrates a second useful property of levers, namely that levers can be used to amplify an applied force. However, force amplification does not come without cost. If the effort arm is moved down a distance of s_E, the load arm will move up a distance of only:

$$s_L = \frac{r_L}{r_E} \cdot s_E = \frac{s_E}{\text{IMA}}$$

and we will have $s_L < s_E$ whenever IMA > 1. Increased force comes at the expense of decreased movement. This inverse relationship between movement and force is not always a disadvantage. Consider the lever with $r_E < r_L$. In this case:

$$\frac{1}{\text{IMA}} = \frac{r_L}{r_E} > 1$$

Therefore the distance s_L moved by the load arm will be greater than the distance s_E moved by the effort arm. This illustrates a third useful property of levers, namely that levers can be used to amplify motion.

A load does not always have to be a mass that is being pulled downward by gravity. Consider the case shown in Fig. 1-23, in which the load is a spring pulling up on the load arm, while the effort is applied in an upward direction so as to stretch the spring and lower the tip of the load arm.

Second and third class levers

So far we have considered only levers in which the fulcrum lies between the load and the applied effort. Two other configurations are possible. In a *second class*

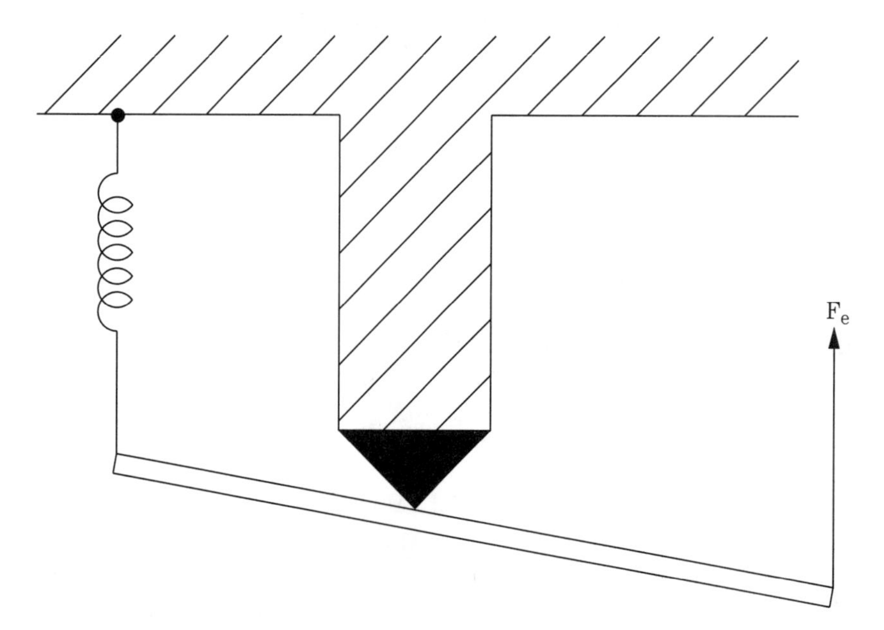

1-23 First class lever with fulcrum above the lever.

lever, the load lies between the fulcrum and the applied effort, as shown in Fig. 1-24. The definitions for IMA and AMA are the same as for a first class lever. However, by definition, r_E will always be greater than r_L, so the IMA will always be greater than 1. In a *third class lever*, the effort is applied at a point lying between the fulcrum and the load arm as shown in Fig. 1-25. In this case the effort arm is always shorter than the load arm, so the IMA will always be less than 1. Third class levers are primarily used as movement amplifiers.

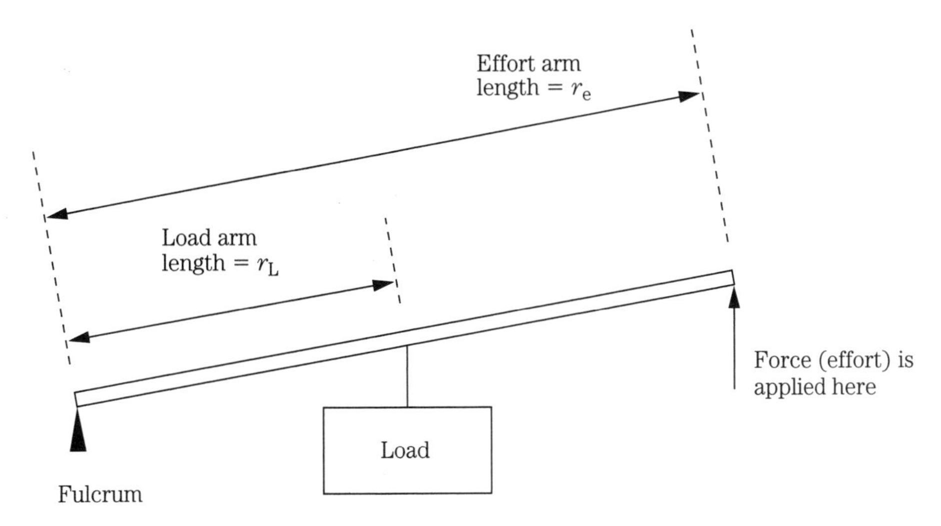

1-24 Second class lever.

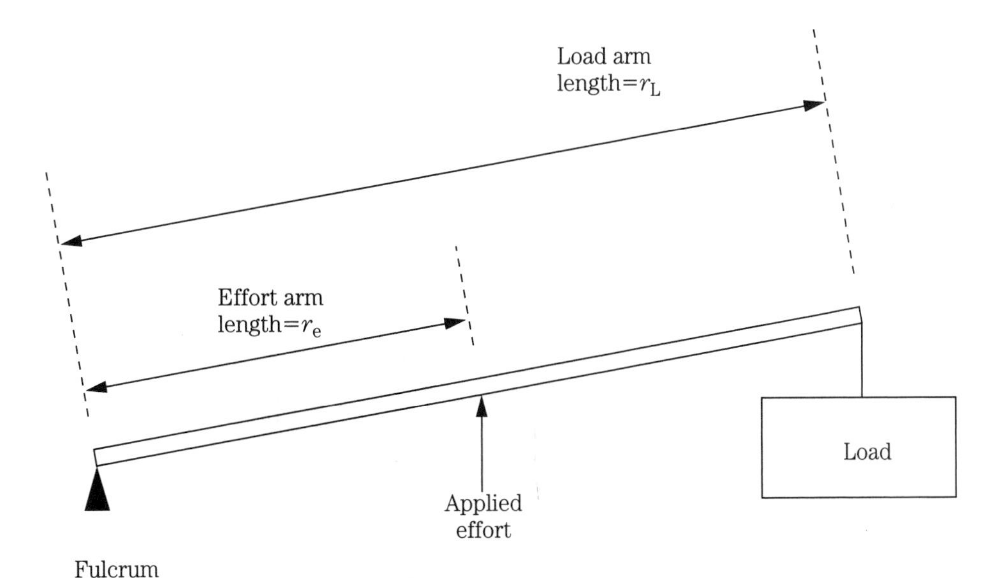

1-25 Third class lever.

Wheels and pulleys

A wheel and axle can be thought of as a generalization of the lever. Consider the second class lever shown in Fig. 1-26.

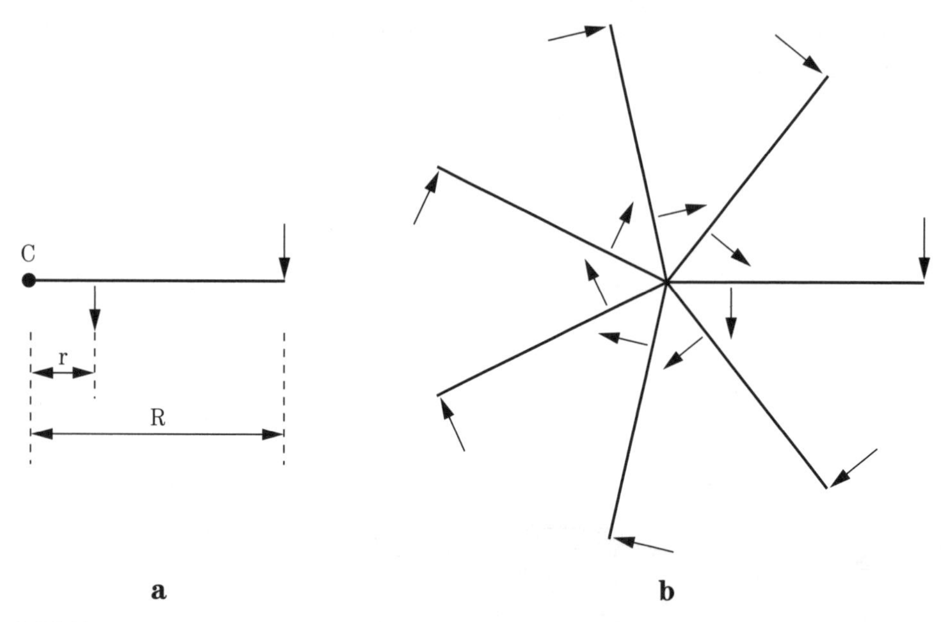

1-26 Many levers around a common fulcrum approximate a wheel and axle.

The fulcrum is at **C**, the input arm has length R, and the output arm has length r. The TMA of this lever is R/r. We could make many copies of this lever positioned at various angles around the common fulcrum point as shown in Fig. 1-26b. In the limit, as the number of levers approaches infinity and the angle between adjacent levers approaches zero, we obtain the wheel and axle combination shown in Fig. 1-27. The wheel is rigidly attached to the axle. The fulcrum lies at the center line of the axle, the output force is applied tangentially to the axle, and the input force is applied tangentially to the wheel rim. The TMA remains R/r.

Suppose we take the wheel-and-axle combination from Fig. 1-27 and fasten a rope to the axle as shown in Fig. 1-28. The result is a simple windlass. Turning the wheel will cause the rope to be wound up onto the axle, with a theoretical mechanical advantage of R/r.

Figure 1-29 shows a modified wheel commonly called a pulley. The entire assembly shown in Fig. 1-29 is called a *block*, and the pulley is more properly referred to as the *sheave*. A rope can be passed around the pulley and used to lift a load as shown in Fig. 1-30. This configuration merely changes the direction of the applied force. The amount of force required or the distance through which the force must act both remain unchanged, so the TMA is 1.

It is possible to gain mechanical advantage with pulleys if they are used correctly. The configuration shown in Fig. 1-31 has a TMA of 2. For each foot of rope

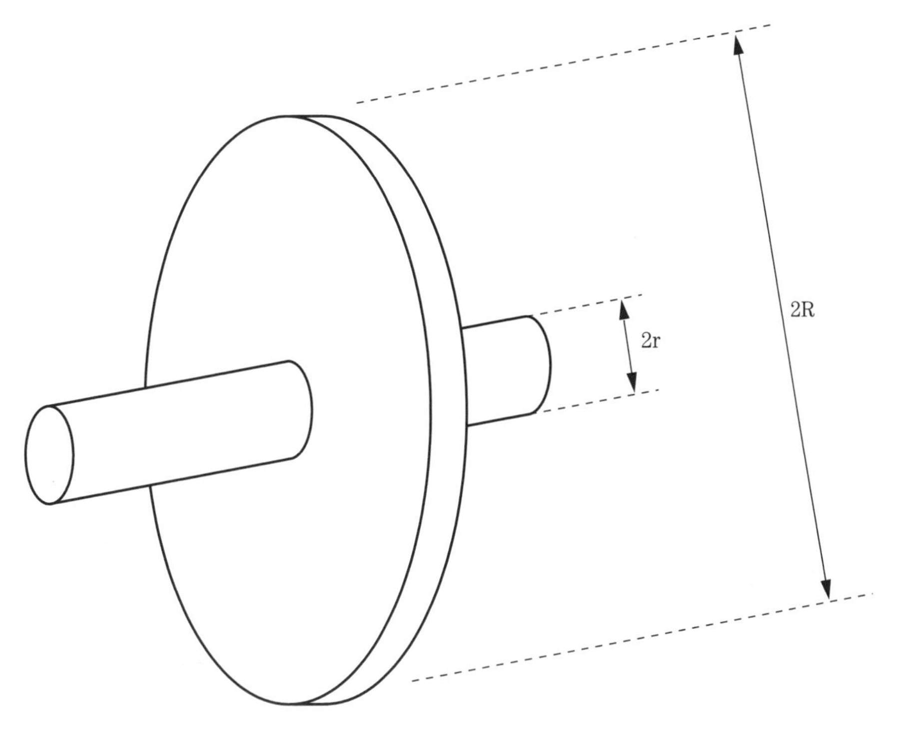

1-27 Wheel and axle.

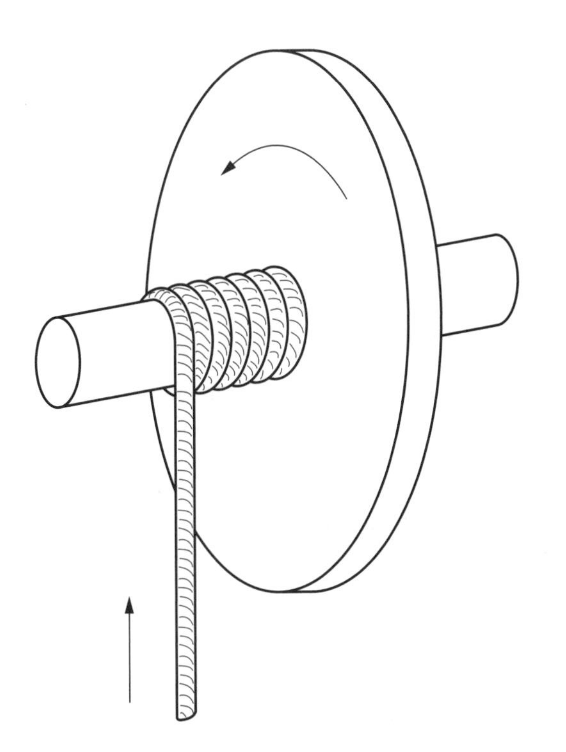

1-28
A wheel and axle used as a windlass.

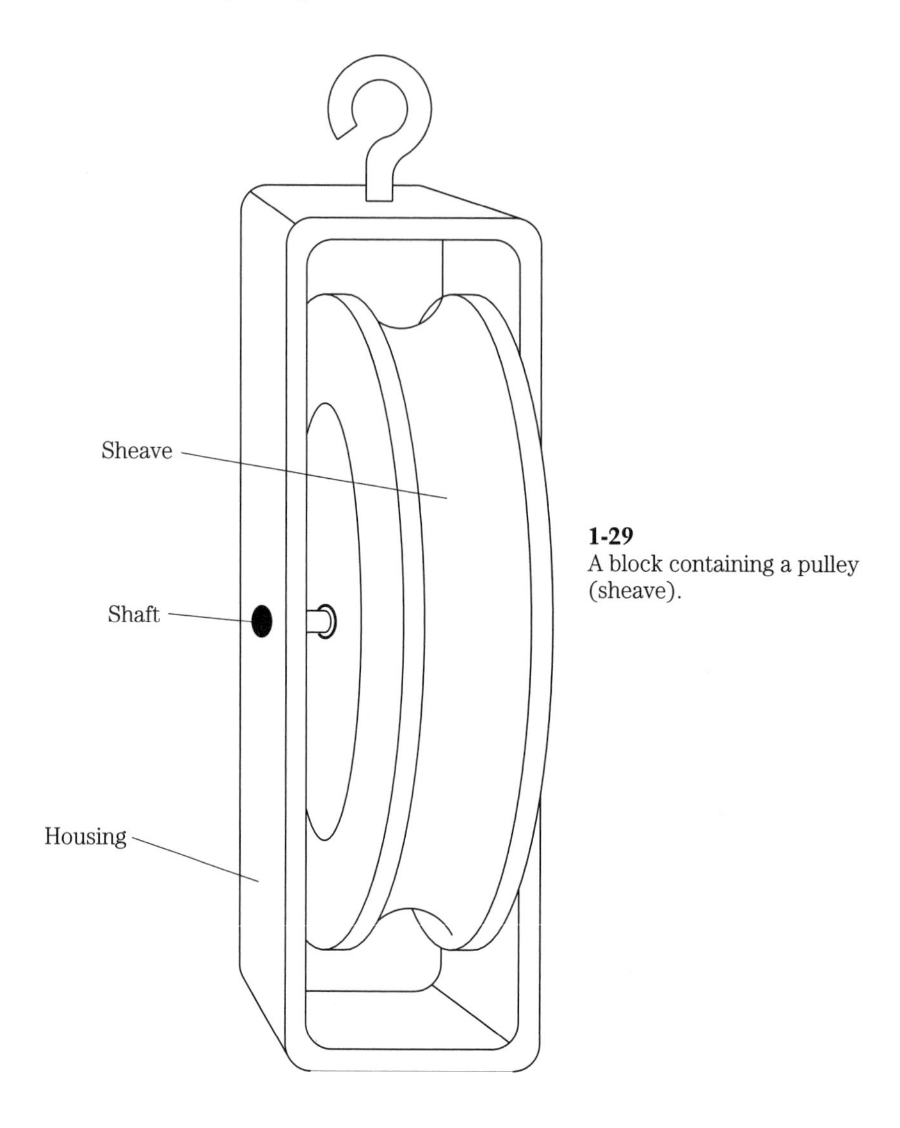

1-29
A block containing a pulley (sheave).

pulled through the top pulley, strands **A** and **B** will each be shortened by half a foot, lifting the load by half a foot.

In general, the TMA is equal to the number of strands supporting the load. The final strand (such as strand **C** in Fig. 1-31), which serves to reverse the direction of the applied force, does not support the load and does not count in figuring the TMA. However, in a configuration such as Fig. 1-32, the final strand does support the load and does count in figuring the TMA. The configuration of Fig. 1-32 has a theoretical mechanical advantage of 3. Additional pulley configurations are discussed in chapter 7.

Inclined planes and screws

Anyone who has ever tried to load a heavy lawnmower into the back of a pickup truck already appreciates the usefulness of the inclined plane. The inclined plane

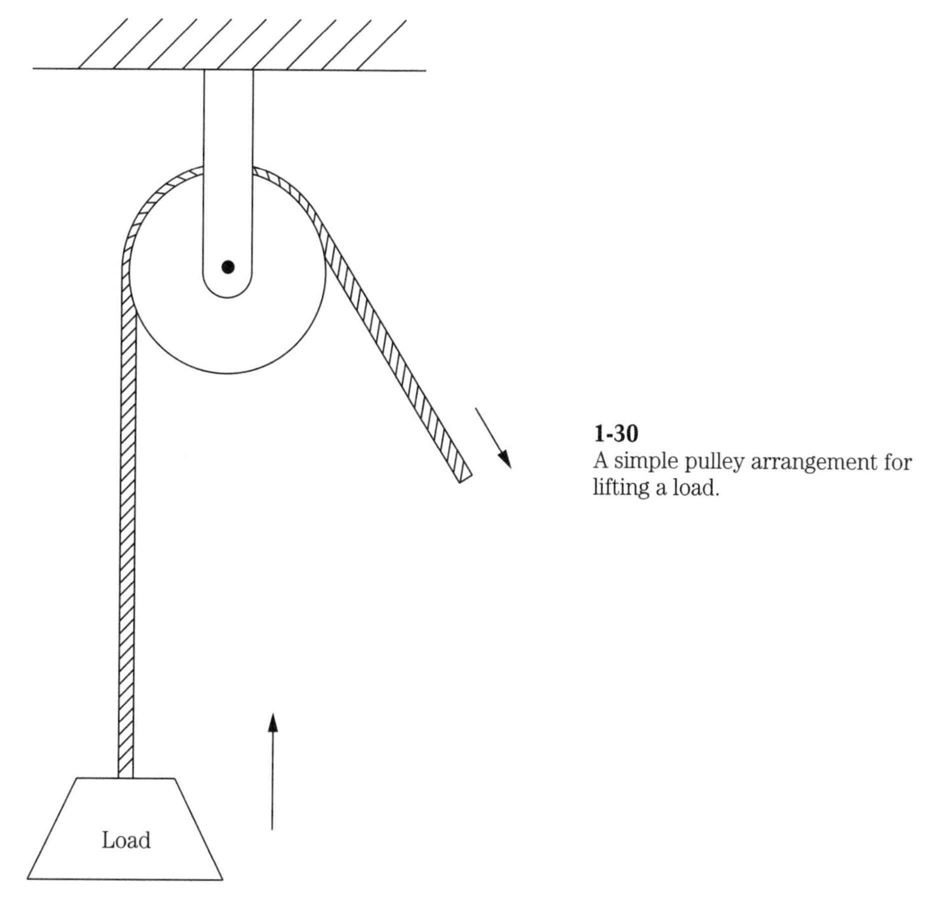

1-30
A simple pulley arrangement for lifting a load.

shown in Fig. 1-33 allows the load to be raised a distance of y by traversing the sloping face that has a length of r, which can be obtained as follows:

$$r = \sqrt{x^2 + y^2}$$

The theoretical mechanical advantage is then r/y. Unlike levers and pulleys in which the TMA is often a reasonable approximation to the actual mechanical advantage (AMA), inclined planes often exhibit significant frictional losses which can make the AMA very much less than the TMA.

Screws

Take an inclined plane and wrap it around the surface of a cylinder. The result is a screw. Aside from their usual applications as fasteners, the most likely use for screws in amateur mechanical applications is to convert rotary motion into linear motion.

Figure 1-34 illustrates how a section of threaded rod and mating nut can be used to convert the rotary motion from a motor into linear motion. An extra hole is drilled in the nut to accommodate a smooth rod, or perhaps a section of piano wire. As the threaded rod is rotated, the piano wire prevents the nut from turning, and thereby forces the nut to advance along the length of the threaded rod.

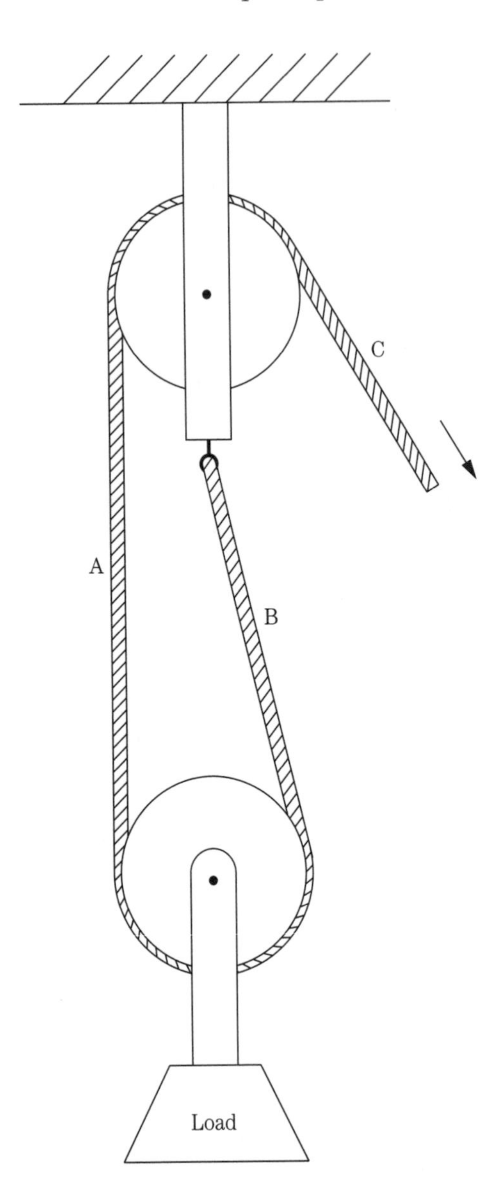

1-31
A pulley arrangement with a theoretical mechanical advantage of 2.

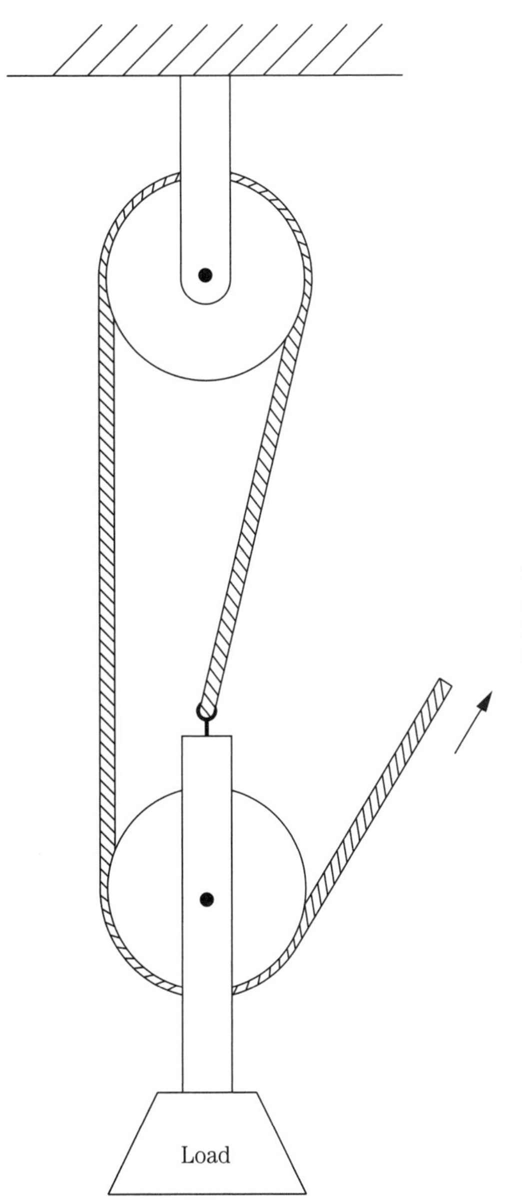

1-32
A 2-pulley arrangement with a theoretical mechanical advantage of 3.

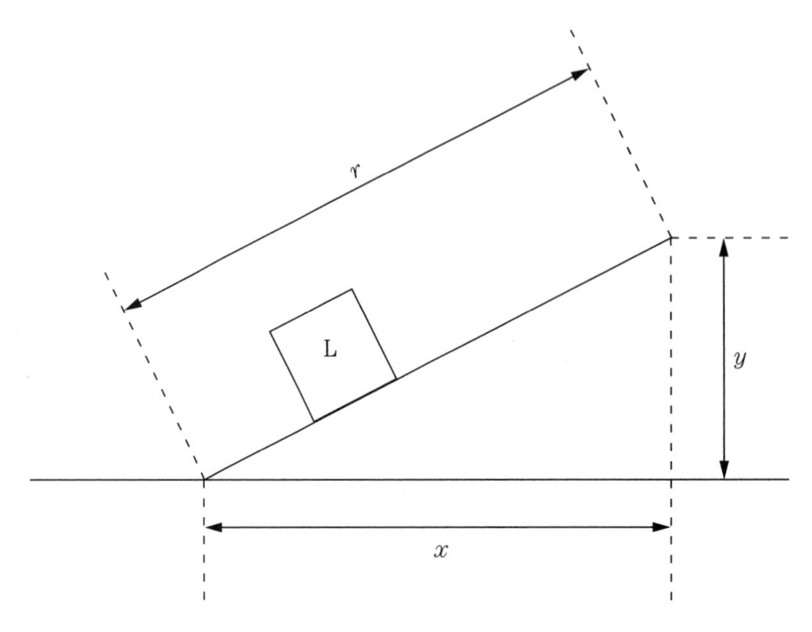

1-33 An inclined plane.

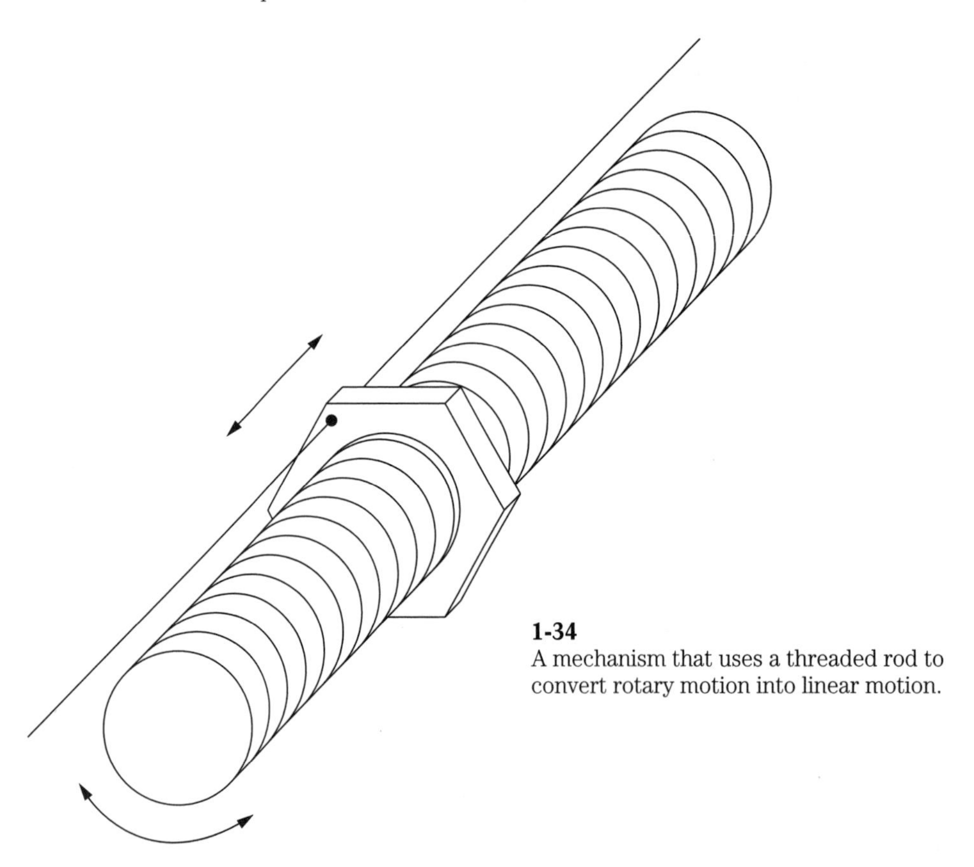

1-34
A mechanism that uses a threaded rod to convert rotary motion into linear motion.

2
CHAPTER

Sensors and control

A robot or other experimental electromechanical system will typically contain an assortment of motors, solenoids, positioners, and actuators in addition to an even more varied assortment of sensors. The control system is that portion of the overall system that is concerned with causing the system to accurately and dependably do the operator's bidding. In order to have the complete system perform as desired, it may be necessary to control a number of the system's physical parameters such as speed, direction, angular position, voltage, current, or fluid pressure. Control is a big deal; a comprehensive introduction to control theory could easily fill two books the size of this one. This chapter is not meant to cover all of this material in even a very superficial way. Rather, the intent is to expose a few fundamental ideas of control and then discuss how control considerations impact design, selection, and use of electrical or mechanical sensing and positioning components.

Open-loop control

Some things, such as the voltage supplied to a motor, can be controlled directly; while other things, such as motor speed, can only be controlled indirectly via control of things that have an impact on motor speed; voltages, currents, etc. However, in many cases, if we attempt to control a motor's speed by controlling its electrical inputs, we may have only limited success in achieving the desired speed. The mechanical load on the motor may be variable and unknown, but this load will have an impact on the motor's speed. A simple circuit for adjusting the speed of a permanent magnet motor is shown in Fig. 2-1. From a control system point of view, this circuit can be represented by the block diagram shown in Fig. 2-2. This is known as an *open-loop* system because the control system makes changes in the *actuating signal* (armature voltage) in an attempt to achieve some desired value of the *controlled variable* (speed) without making use of any feedback regarding how the changes made are actually affecting the controlled variable. Open-loop control has advantages and disadvantages. It is simple to build and requires no sensors, but it is prone to errors caused either by unknowns in the process model or by unobserved inputs to the process.

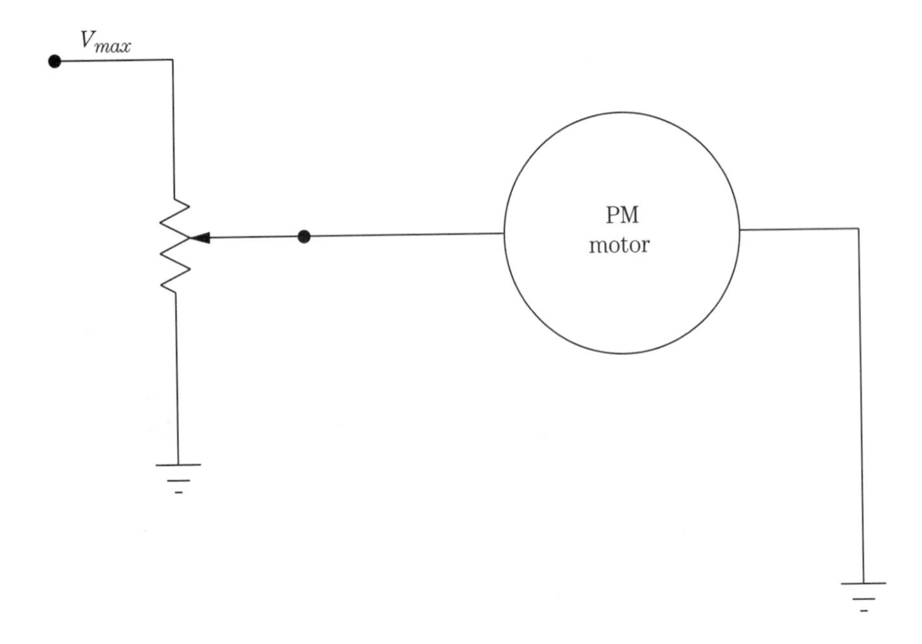

2-1 Simple speed control for a PM motor.

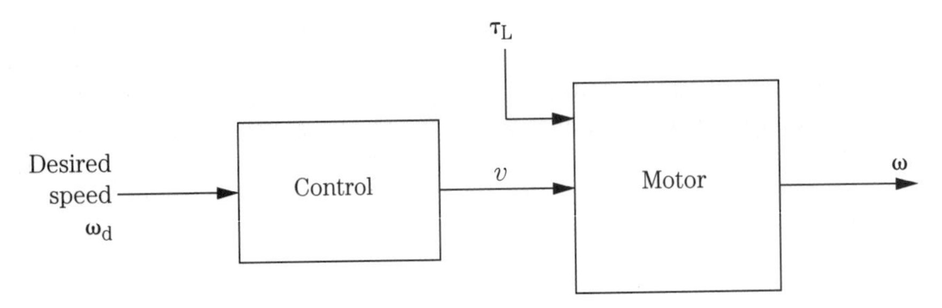

2-2 Block diagram of open-loop system for controlling motor speed.

In open-loop control systems, a piece of hardware is commanded to do something and "trusted" to do whatever it is told. For example, a robot may be commanded to roll forward for a certain interval of time, which based on the wheel diameters and assumed speed of rotation should correspond to a certain distance. Suppose our intent was to move the robot forward a distance of five feet, the wheels have a circumference of 12 inches, and they rotate at a speed of 30 rpm. Consequently, we command the motor to drive the wheels in a forward direction for 10 seconds. Unfortunately, things in the real world have a way of messing up the best of plans. Perhaps the robot is running in deep carpet laid over thick padding, and the extra torque required to turn the wheels causes the motor to run slower than normal. In this case, running the drive motor in a forward direction for 10 seconds will move the robot forward somewhat less than the intended five feet. The solution to this problem lies in some form of closed-loop control.

The speed-control problem

The problem of controlling a motor's speed can take on a number of different subtle variations.

Coarse control

In a toy racing car, the control can be relatively simple; the "driver" causes the motor to slow down or speed up as desired without regard to the actual speed of the motor or car. Relatively crude gradations such as very slow, slow, medium, fast, and very fast are sufficient. If the load on the motor, and consequently the car's actual speed, varies somewhat due to changing track conditions, it is not a big deal. Severe long-term changes (such as driving off the path into the grass) can be accommodated by the driver's observing the change and exercising the appropriate speed-up or slow-down control maneuver.

Regulation

In constant speed control, the motor's speed is held to some constant (but unknown) value over a wide variation in load conditions. This type of control is also called *regulation*. It is doubtful that this type of speed control would ever be selected for a robot's travel within its environment. However, this type of control could be used for some robotic functions; perhaps an arm that moves up or down at a constant rate regardless of the load being lifted by the arm. In many cases, regulation is accomplished via careful electronic adjustment of the driving voltage to compensate for expected variations in the voltage-to-speed relationship of a particular motor.

Fixed speed

In fixed speed control, the motor speed is held to some particular value over a wide variation in load. For some practical control schemes it may be important to know that the gripping pinchers of a robotic arm close at some particular rate such as 0.2 inch per second. In such cases, fixed speed motor control may be called for. Usually this type of control will require that the controlled variable be measured and that these measurements be fed back to cause the appropriate correction in the actuating signal. A system of this type is called a *closed-loop* control system.

Closed-loop control

In contrast to open-loop systems, closed-loop control systems continuously measure the controlled variables and compare the measured values to the desired values. The control system takes action to eliminate or reduce the observed error between the measured and desired values of the controlled variables. A block diagram of a motor speed controller incorporating feedback is shown in Fig. 2-3. Closed-loop control can be applied to the robot situation described above. A tachometer can be used to sense the rotational speed of the wheels, and the run time adjusted as required to ensure that the robot advances by the desired distance. The signal from

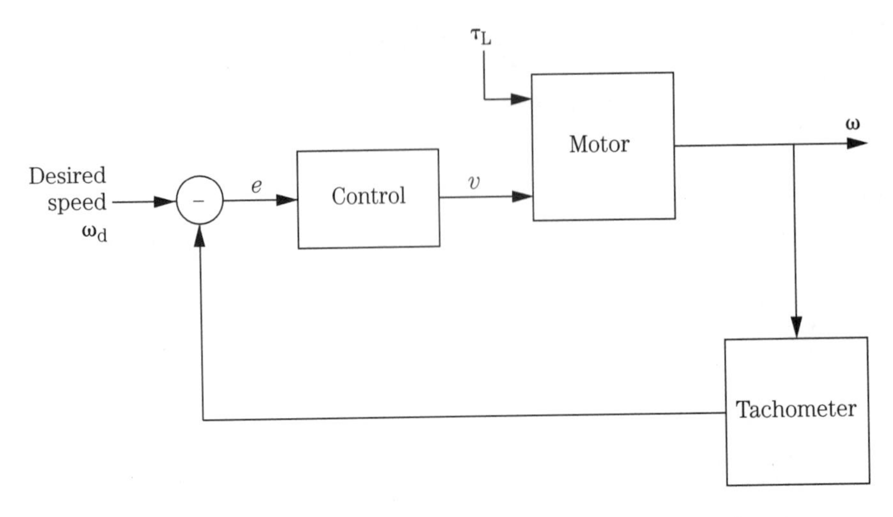

2-3 Block diagram of a closed-loop system for controlling motor speed.

the tachometer is referred to as feedback. Sometimes it is not so easy to obtain an appropriate feedback signal for use in a closed-loop control system. Staying with our robot example, suppose now that instead of being bogged down in thick carpeting, the robot is attempting to roll up a tiled incline. In this case, the wheels will possibly slip and each rotation of the wheels will advance the robot by something less than 12 inches. The amount of slippage can vary, and it is not an easy thing to measure in a way that can be used as feedback to the drive control. Closed-loop control systems in robotic applications frequently need to make measurements of mechanical quantities such as speed, position, and force.

Sensing rotational speed

A device for measuring rotational speed is called a *tachometer*. There are several different ways that a tachometer can be implemented in the home shop.

dc generator

If the shaft of a dc permanent-magnet motor is mechanically rotated, a pulsing voltage like the one in Fig. 2-4 can be measured across the motor's terminals. The cheap PM motors found in toys will usually have a 2-segment commutator and will produce two pulses per shaft revolution. Many of the newer digital multimeters (DMMs) are capable of measuring frequencies up to several hundreds of kilohertz. Often, such a meter can be used directly to measure the fundamental frequency of the pulse train output from the motor. This approach may not lend itself to being "built into" a robot, but it can be very useful for assessing the characteristics of undocumented motors.

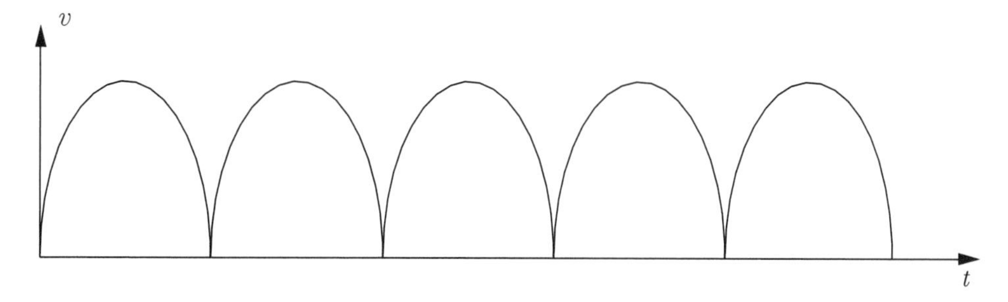

2-4 Pulsed voltage output from a mechanically driven PM motor.

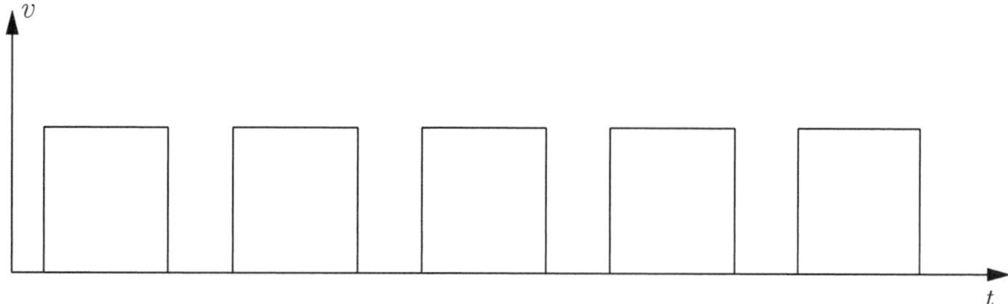

2-5 Motor-output pulses after being "squared up" by a comparator circuit.

An alternative approach uses comparator circuits to "square up" the pulses of Fig. 2-4 to obtain pulses similar to those in Fig. 2-5. At typical motor speeds, it is a straightforward matter to design a simple logic circuit to count the number of such pulses that occur in a fixed time interval. This count will be directly proportional to the shaft speed of the motor.

This approach has several drawbacks associated with the use of a second motor for speed sensing:

- The "sensing" motor must be coupled to the drive motor whose speed is being measured. This will introduce additional load on the driven motor.
- There must be space for the sensing motor. Many inexpensive motors only have useable shafts extending from one end, so the sensing motor and the useful load must both be somehow coupled to the same end of the shaft. Sometimes the space inside mechanical assemblies gets very cramped!

Other approaches

Two additional approaches for measuring motor speed are shown in Fig. 2-6. In one approach, a disc of opaque material is fixed to the shaft. A hole in this disc allows light from a lamp or LED to reach a photodiode for a short interval once each revolution. This will cause a pulse in the photodiode's output. The other approach is similar, but substitutes a magnet and Hall effect switch for the light source and photodiode.

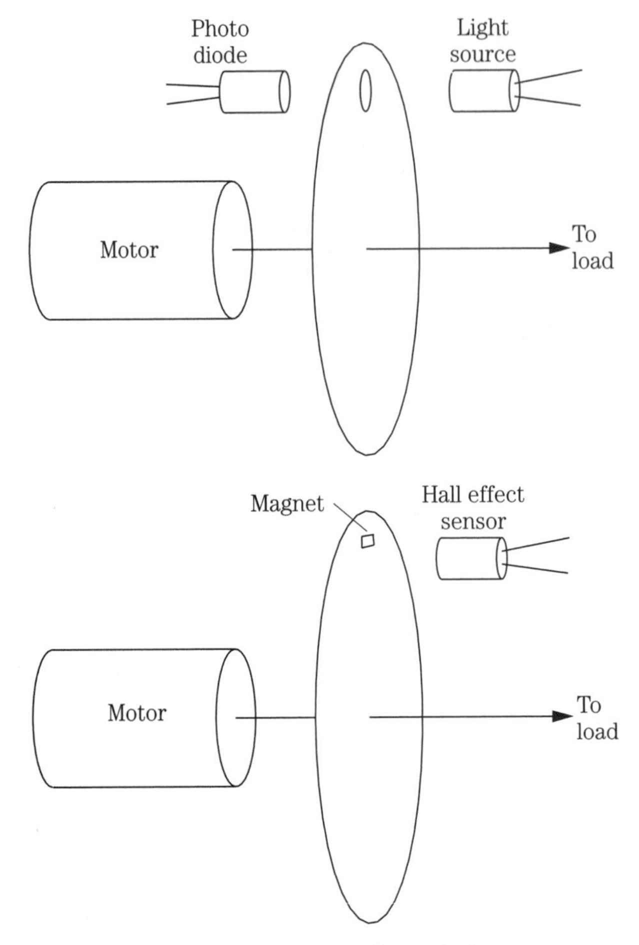

2-6 Schemes for sensing rotational speed of a motor.

Position sensing

Position sensing really spans a wide range of possible situations. Sensing a free-roving robot's position within a building could be considered a "high end" application, while sensing the angle of a finger joint in a gripper would be more of a "low end" application. The "high end" problem of sensing a robot's location is very different from the "low end" problem of sensing the relative position of components within some mechanism.

Angular position sensor #1

A simple position sensor can be constructed from an ordinary potentiometer. In very low stress applications, it may be possible to use the potentiometer itself as the joint pivot. In most applications however, such an arrangement will not be practical. Figure 2-7 shows how two gears can be used to drive a potentiometer that is not coaxial with the pivot. The large gear must be centered on the pivot. In this example

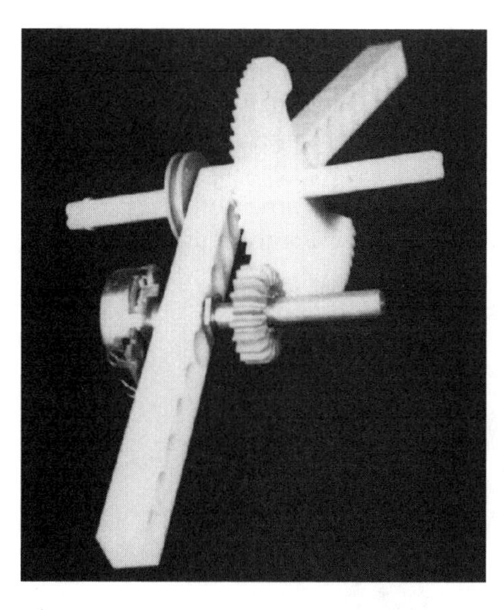

2-7
Two gears drive a potentiometer used as a position sensor for a revolute joint.

mechanism constructed from Gears-In-Motion parts, the large gear is actually mounted on the pivot shaft as a matter of convenience. In a custom design, the center portion of the gear can be removed as necessary to provide clearance for a more sophisticated pivot. Some teeth on this large gear will never be in a position to mesh with the small gear, so they have been cut away.

Angular position sensor #2

The potentiometer-based sensor described above is limited by the rotational range of the potentiometer. If unlimited freedom of rotation is required, an *optical encoder* can be used to sense angular position. Optical encoders consist of a light source, a light detector, and a code disc as shown in Fig. 2-8. As the shaft rotates, opaque and transparent sections alternately pass between the light source and detector producing a series of pulses on the detector output. Either a general purpose processor or a dedicated counter circuit can be used to count the pulses and determine the number of wedges that the shaft has rotated. Obviously, the narrower the wedges, the more precisely partial rotations can be metered.

2-8
A relative optical encoder disc.

An *absolute encoder* uses multiple detectors and a different type of code wheel to indicate shaft position directly without a need to count pulses. Figure 2-9 shows a code wheel for an absolute encoder capable of resolving shaft positions to within 45°. This code wheel would require three different light detectors and the outputs of these detectors would provide binary coded values 0, 1, 2, 3, 4, 5, 6, or 7 depending upon the angular position of the disc. Low-to-medium precision discs can be easily fabricated in the home shop using clear plastic and black paint or fingernail polish.

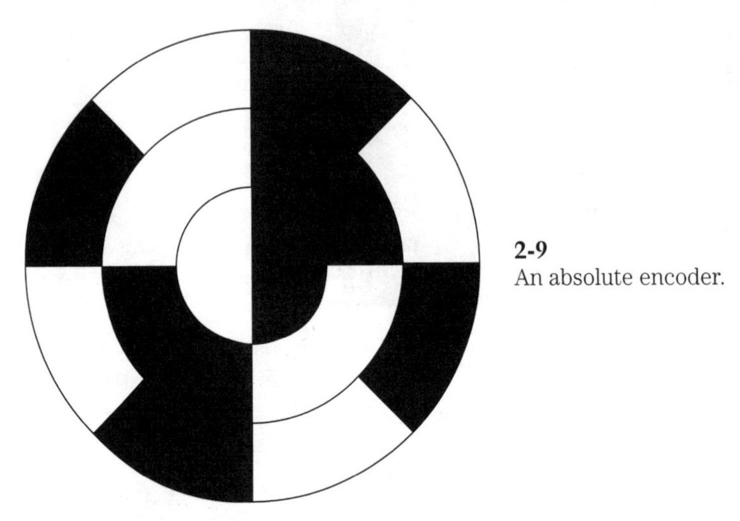

2-9
An absolute encoder.

Translational position

Translational position sensors can be designed as linear analogs of the angular sensors already discussed. The geared potentiometer schemes can be extended to linear motion by replacing the gear on the moving arm with a rack. Long racks are often difficult for experimenters to get hold of, so it may be advantageous to replace the gear with a rubber wheel that makes direct contact with the sliding arm as shown in Fig. 2-10.

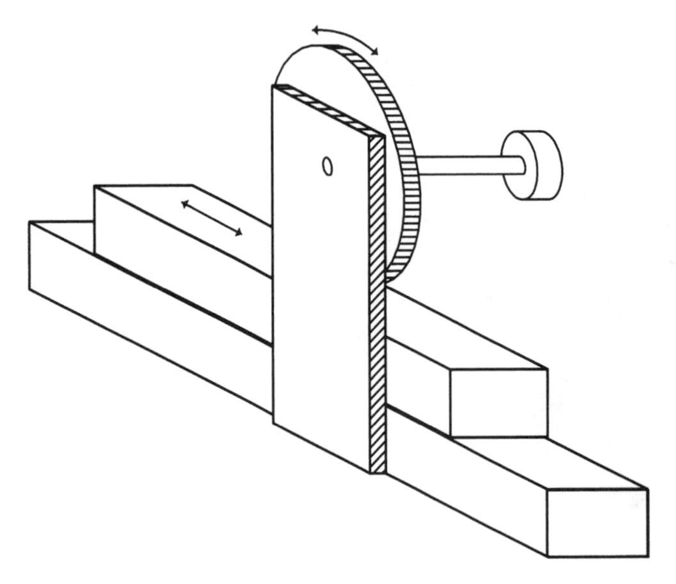

2-10
Friction wheel in contact with sliding member turns potentiometer to indicate position.

3

CHAPTER

Motors

There is one basic idea that is the key to understanding the electrical behavior of motors:

When a motor is operating as a motor, it is also acting a bit like a generator.

Just what does this mean? When a motor is running, there will be an **internally generated** voltage E_A in the armature that is caused by the coils of the armature moving within the motor's magnetic field. This internally generated voltage should not be confused with the external voltage that is applied to the armature. (See Fig. 3-1.) The voltage E_A is a function of the magnetic flux ϕ in the motor and the speed ω of the rotor:

$$E_A = k\phi\omega \tag{3-1}$$

where k is a constant based on details of the motor's construction. In a motor, the polarity of E_A will always oppose the externally applied voltage so the KVL equation for the armature circuit is written as:

$$V_T = E_A + I_A R_A \tag{3-2}$$

where

V_T = voltage applied to external terminals
I_A = armature current
R_A = armature resistance

Contrast this with the case of dc generators in which the polarity of E_A will agree with the externally applied voltage and for which the KVL equation for the armature circuit is written as:

$$V_T = I_A R_A - E_A$$

The armature current determines the amount of torque τ_{ind} induced upon the rotor:

$$\tau_{\text{ind}} = k\phi I_A \tag{3-3}$$

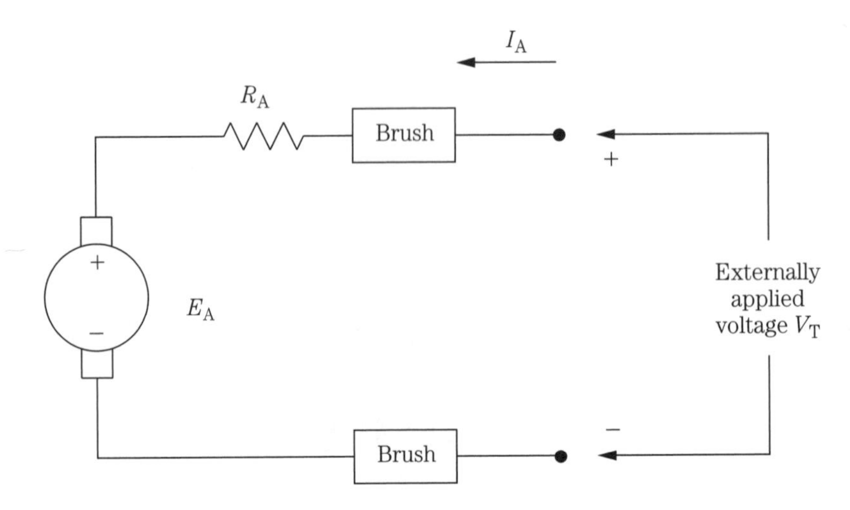

3-1 Equivalent circuit for KVL analysis of a motor's armature circuit.

We can combine Eqs. 3-1, 3-2, and 3-3 to obtain:

$$\omega = \frac{V_T}{\phi k} - \frac{R_A}{(\phi k)^2}\,\tau_{ind} \tag{3-4}$$

Based on Eq. 3-4, the speed versus torque characteristic is a straight line (as shown in Fig. 3-2) having a vertical intercept of $V_T/(\phi k)$ and a slope of $-R_A/(\phi k)^2$. Equation 3-4 is somewhat of an idealization. In the real world, many motors will exhibit a phenomenon called *armature reaction* that tends to weaken the motor flux as the load increases. This armature reaction will cause the speed versus torque characteristic to curve upward as shown in Fig. 3-3. This effect is less pronounced in permanent magnet motors than it is in wound-field motors.

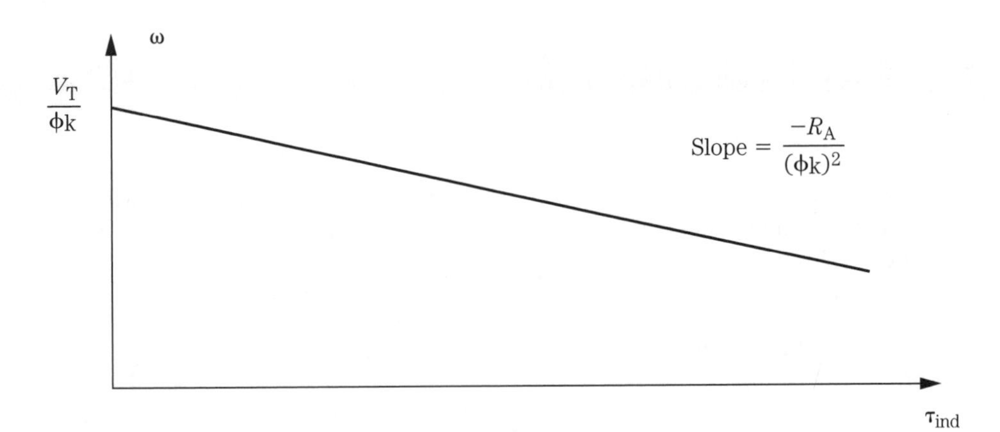

3-2 Idealized speed-versus-torque characteristic for a dc motor.

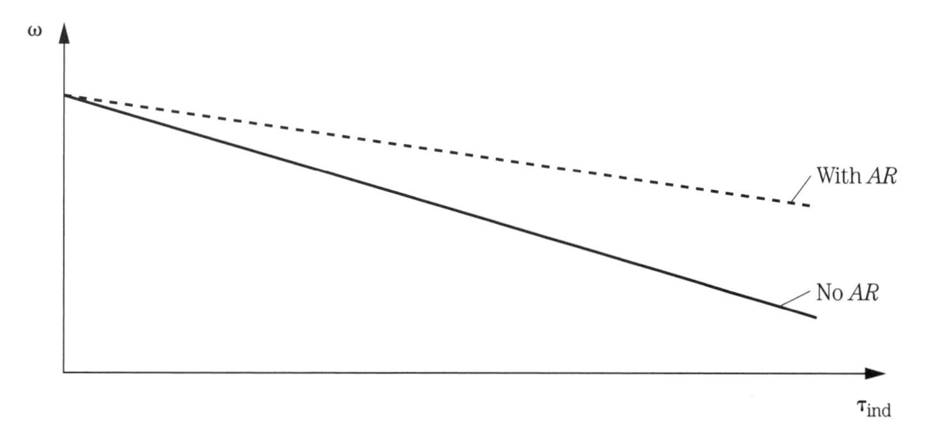

3-3 Speed-versus-torque characteristic for a dc motor.

Magnetic flux is one of the parameters used to characterize a motor's behavior. In a wound-field motor, the flux is a function of the field current I_F as shown in Fig. 3-4. As stated in Eq. 3-1, the internally generated armature voltage E_A is a function of the magnetic flux ϕ and speed ω. Instead of the plot of ϕ versus I_F shown in Fig. 3-4, it is customary to present the magnetization curve as shown in Fig. 3-5 that plots E_A versus I_F for some specified constant speed ω_0.

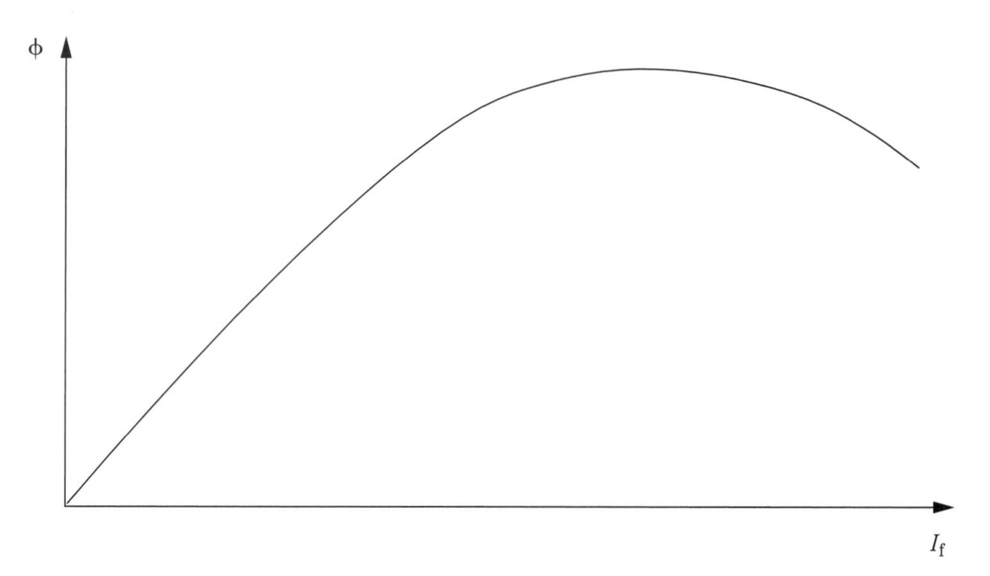

3-4 Flux as a function of field current for a typical motor.

Permanent-magnet dc motors

Permanent-magnet motors are so named because they use permanent magnets instead of field windings to create the field flux for the motor. Toys and other inexpensive items often contain permanent-magnet dc motors. Therefore, this type of

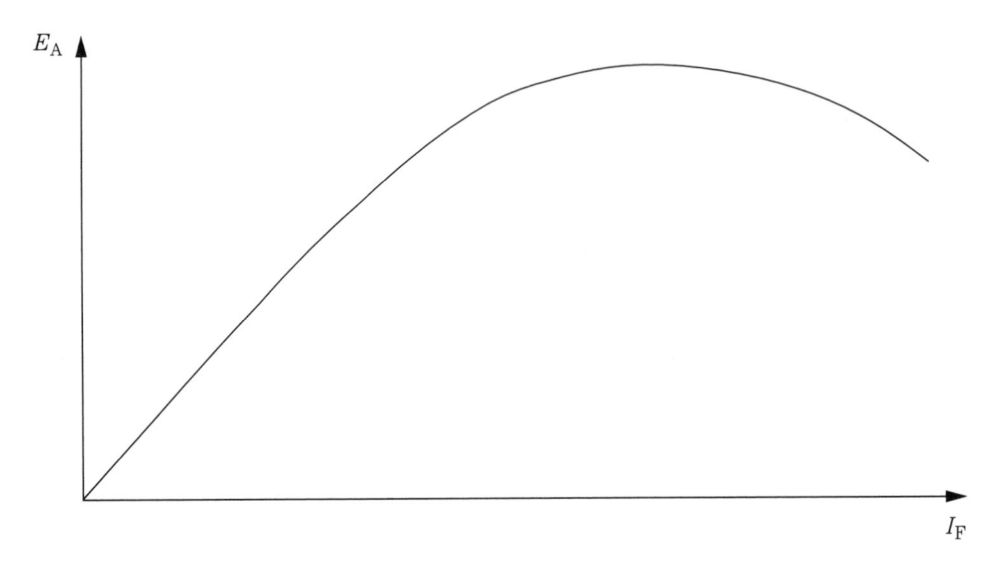

3-5 Magnetization curve for a typical motor.

motor is one of the easiest for experimenters to obtain. An end-on view of a typical permanent-magnet motor is shown in Fig. 3-6. The *stator* consists of permanent magnets that provide a constant magnetic field. Current is passed to the *armature windings* via *brushes* and *commutator segments*. Connections between the windings and the commutator segments are arranged so as to change the polarity of the armature at the appropriate times for continuous rotation of the armature.

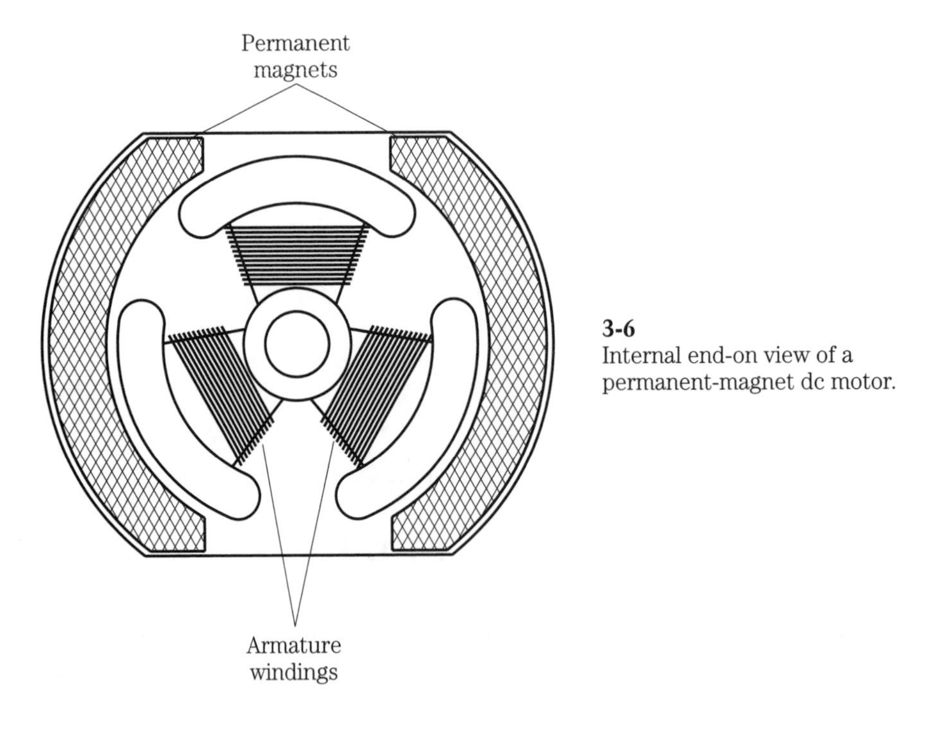

3-6 Internal end-on view of a permanent-magnet dc motor.

Electrical properties

The equivalent circuit of a permanent-magnet dc motor is shown in Fig. 3-1. As discussed earlier in this chapter, if the voltage drop across the brushes is neglected, the equations that govern the motor's behavior are:

$$E_A = k\phi\omega \tag{3-5}$$

$$V_T = E_A + I_A R_A \tag{3-6}$$

$$\tau_{\text{ind}} = k\phi I_A \tag{3-7}$$

where

E_A = internally generated armature voltage
k = a constant based on details of the motor's construction
ϕ = magnetic flux
ω = rotational speed of the motor
V_T = voltage applied to the motor's external terminals
I_A = armature current
R_A = armature resistance
τ_{ind} = induced torque

Equations 3-5, 3-6, and 3-7 can be combined to yield the speed versus torque relationship:

$$\omega = \frac{V_T}{\phi_k} - \frac{R_A}{(\phi k)^2}\tau_{\text{ind}} \tag{3-8}$$

As shown in Fig. 3-2, Eq. 3-8 is the equation of a straight line having a vertical intercept of $V_T/(\phi k)$ and a slope of $-R_A/(\phi k)^2$.

Of all the parameters listed above, the applied voltage V_T is the only one over which we can exercise direct control. What can we say about the motor's reaction to a change in armature voltage? If V_T is **increased**, the following behavior will be observed:

- Based on Eq. 3-6, an increase in V_T will cause I_A to increase.
- Based on Eq. 3-7, an increase in I_A will cause an increase in the induced torque τ_{ind}.
- When τ_{ind} increases above τ_{load}, the motor will increase its speed ω.
- Based on Eq. 3-5, an increase in ω will cause the internally generated voltage E_A to increase.
- Based on Eq. 3-6, increasing E_A causes the armature current I_A to decrease.
- The current I_A will continue decreasing, causing τ_{ind} to decrease until $\tau_{\text{ind}} = \tau_{\text{load}}$, thus causing the speed to stop increasing.

Another way to look at the impact of increasing V_T involves the speed versus torque relationship given by Eq. 3-8. As shown in Fig. 3-7, an increase in V_T causes the vertical intercept (no-load speed) to increase, while keeping the slope constant.

Reaction to changes in load

How will a permanent-magnet motor react to changes in mechanical load? If the applied load increases while the terminal voltage V_T is held constant, the following behavior will be observed:

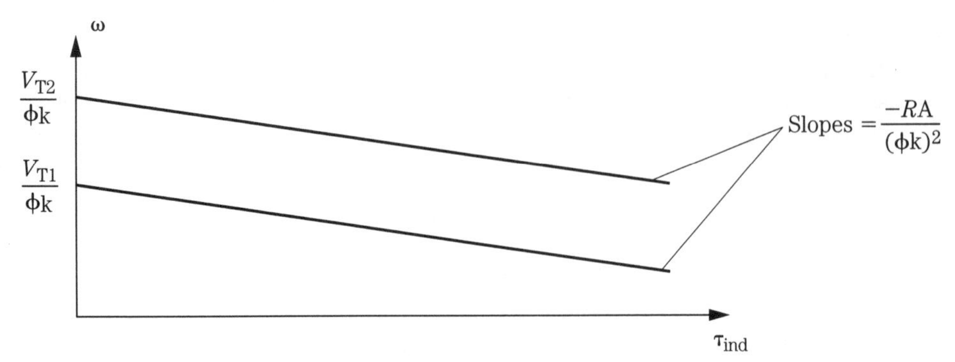

3-7 Speed-versus-torque characteristics of a permanent-magnet motor for two different loads.

- Because $\tau_{load} > \tau_{ind}$, the motor will begin to slow down. Based on Eq. 3-5, a decrease in speed ω will cause a decrease in the internally generated voltage E_A.
- Based on Eq. 3-5, a decrease in E_A will cause an increase in the armature current I_A.
- Based on Eq. 3-7, this increase in I_A will cause an increase in the induced torque τ_{ind}. The increase in τ_{ind} will exactly balance the increase in load, but the motor will be operating at a lower speed and drawing increased current.

This behavior is consistent with Eq. 3-8, and the speed versus torque characteristics shown in Figs. 3-2 and 3-7. If they are used in applications requiring anything beyond simple relative speed control, permanent-magnet motors will need fairly sophisticated control circuitry to maintain constant or fixed speeds over a range of load torque.

Separately excited dc motors

A separately excited dc motor can be thought of as a generalized permanent-magnet motor in which a *stator* and *field coil* are used instead of permanent magnets for providing the field flux. An end-on view of a typical wound-field motor is shown in Fig. 3-8.

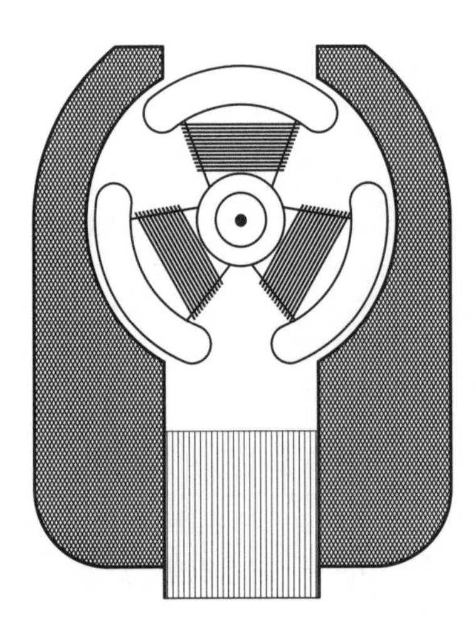

3-8
Internal end-on view of wound-field motor.

Electrical properties

The equivalent circuit of a separately excited dc motor is shown in Fig. 3-9. The equations governing the motor's behavior are similar to the equations for a permanent-magnet motor.

$$E_A = k\phi\omega \tag{3-9}$$

$$V_T = E_A + I_A R_A \tag{3-10}$$

$$\tau_{ind} = k\phi I_A \tag{3-11}$$

$$\omega = \frac{V_T}{\phi_k} - \frac{R_A}{(\phi k)^2}\tau_{ind} \tag{3-12}$$

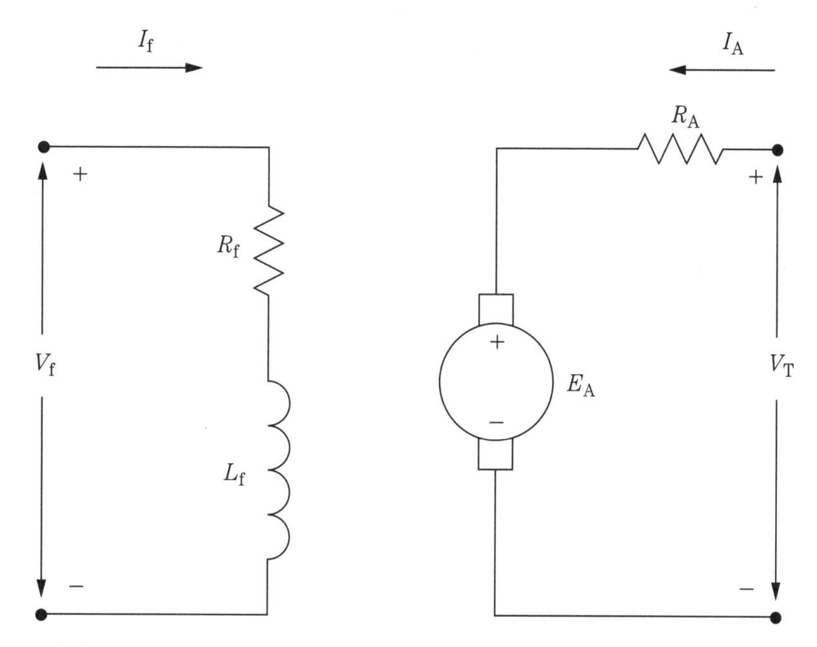

3-9 Equivalent circuit for a separately-excited dc motor.

The major difference is that instead of being constant as it is in permanent-magnet motors, the field flux ϕ is given by:

$$\phi = \frac{\mu N A}{L_C} I_F \tag{3-13}$$

where

μ = permeability of the core material
N = number of turns in the field winding
A = cross sectional area of the core
L_C = mean path length of the magnetic field
I_F = current through the field windings

Actually, for our purposes, μ, N, A, and L_C can be lumped into a single constant to yield:

$$\phi = K_F I_F \tag{3-14}$$

This equation indicates that the flux within a wound-field motor is proportional to the field current. A decrease in I_F causes a proportional decrease in ϕ.

For an increase in V_T, the behavior of a separately excited motor will be the same as the corresponding behavior of a permanent-magnet motor. Unlike a permanent-magnet motor, a separately excited motor can have its flux changed by a change in field current. If ϕ is **decreased,** the following behavior will be observed:

- Based on Eq. 3-9, a decrease in ϕ will cause a decrease in the internally generated voltage E_A.
- Based on Eq. 3-10, a decrease in E_A will cause an increase in armature current I_A.
- Based on Eq. 3-11, an increase in I_A will tend to cause an **increase** in the induced torque τ_{ind}. However, the original decrease in ϕ will tend to cause a **decrease** in τ_{ind}. Which way will τ_{ind} change? In most practical situations, the increase in I_A will be many times larger than the decrease in ϕ, so τ_{ind} will increase.
- As the increase in τ_{ind} makes $\tau_{ind} > \tau_{load}$, the motor will increase its speed ω.
- Based on Eq. 3-9, an increase in ω causes E_A to increase.
- Based on Eq. 3-10, the increase in E_A will cause I_A to decrease.
- The armature current I_A decreases until $\tau_{ind} = \tau_{load}$, and the motor continues to operate at the new increased speed.

Another way to look at the impact of decreasing ϕ involves the speed-versus-torque characteristic given by Eq. 3-12. A decrease in ϕ causes the vertical intercept (no-load speed) to increase, and the slope becomes steeper as shown in Fig. 3-10.

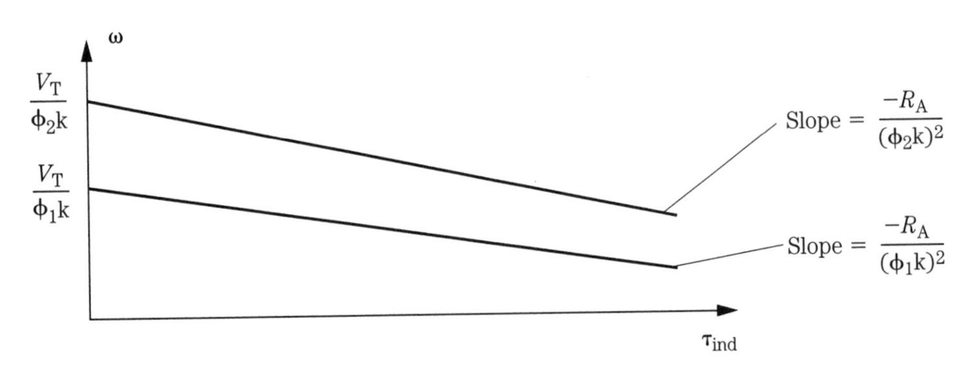

3-10 Speed-versus-torque characteristics of a separately excited dc motor for two different values of field flux.

4
CHAPTER

Motor control

Control of permanent-magnet motors

One major disadvantage of PM (permanent-magnet) motors is the limited number of parameters that can be varied to control the motor's operation. The armature is connected across the motor's only two input terminals. Simple relative speed control is easy; the terminal voltage V_T is increased to cause the motor to speed up, and V_T is decreased to cause the motor to slow down. Such control is conceptually simple, but as we will see in this section, there are good ways and some not-so-good ways to implement the variable voltage.

Maintaining a constant speed over a range of variable torque is not such an easy problem for PM motors. Virtually all control schemes for maintaining constant speed in the face of varying torque requires some sort of feedback. One conceptually straightforward approach involves using a tachometer to measure the motor speed, with the control circuitry arranged to increase V_T if the measured speed is too low and decrease V_T if the measured speed is too high. This approach may be conceptually simple, but the required tachometer represents a significant chunk of hardware. As we will discover, there are conceptually more sophisticated schemes that are easier to implement.

Simple speed control

Many inexpensive toys and model railroad controllers achieve speed control using a variable resistor scheme as shown in Fig. 4-1. Such an approach has several drawbacks. Power is consumed in the variable resistance as well as in the motor. For battery powered applications, this can add up to an unacceptable waste of prime power.

Furthermore, this approach results in even poorer speed regulation than we might at first assume. If V_T were held constant, the speed versus torque characteristic would be as depicted by trace **a** of Fig. 4-2. However, for a fixed position of the control slider on the potentiometer, the voltage V_T does not remain constant. If R_1 is the amount of resistance between the V_{max} terminal and the slider, the voltage V_T will be given by:

$$V_T = V_{max} - R_1 I_1 \tag{4-1}$$

where I_1 is the current flowing through R_1. The current I_1 will vary as I_A varies with load. With increased load, I_1 will increase causing V_T to drop. Decreased V_T will in turn result in reduced speed. The net effect is that the slope of the speed versus torque curve becomes steeper as depicted by trace **b** of Fig. 4-2.

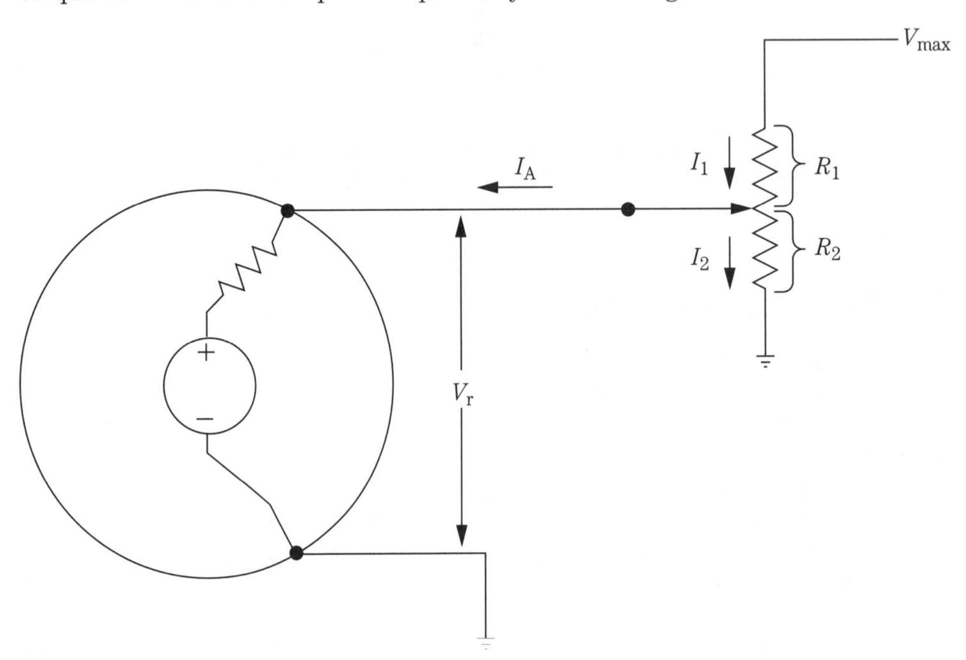

4-1 Simple speed control using a variable resistor.

Silicon-controlled rectifiers

Modern control circuits for PM motors are often built around *silicon-controlled rectifiers* (SCRs). An SCR is schematically depicted as shown in Fig. 4-3. The three terminals are the *anode*, *gate*, and *cathode*. When the voltage applied to the anode is positive relative to the cathode voltage and the gate is unconnected, then negligible current will flow from the anode to the cathode. An SCR connected in this way is said to be *blocked*, *forward-blocked*, or simply *off*. If a positive voltage is then applied to the gate, the SCR will turn *on* allowing a large current to flow from the anode to the cathode.

Once triggered into conduction, the SCR will continue to conduct even if the gate voltage is removed. The SCR will return to the blocked condition if the anode-to-cathode current is forced to drop below some critical level called the *holding current*. This can be accomplished in three ways:

- Limit the current into the anode via resistive or other means.
- Disconnect the anode from its current source.
- Make the anode voltage negative with respect to the cathode.

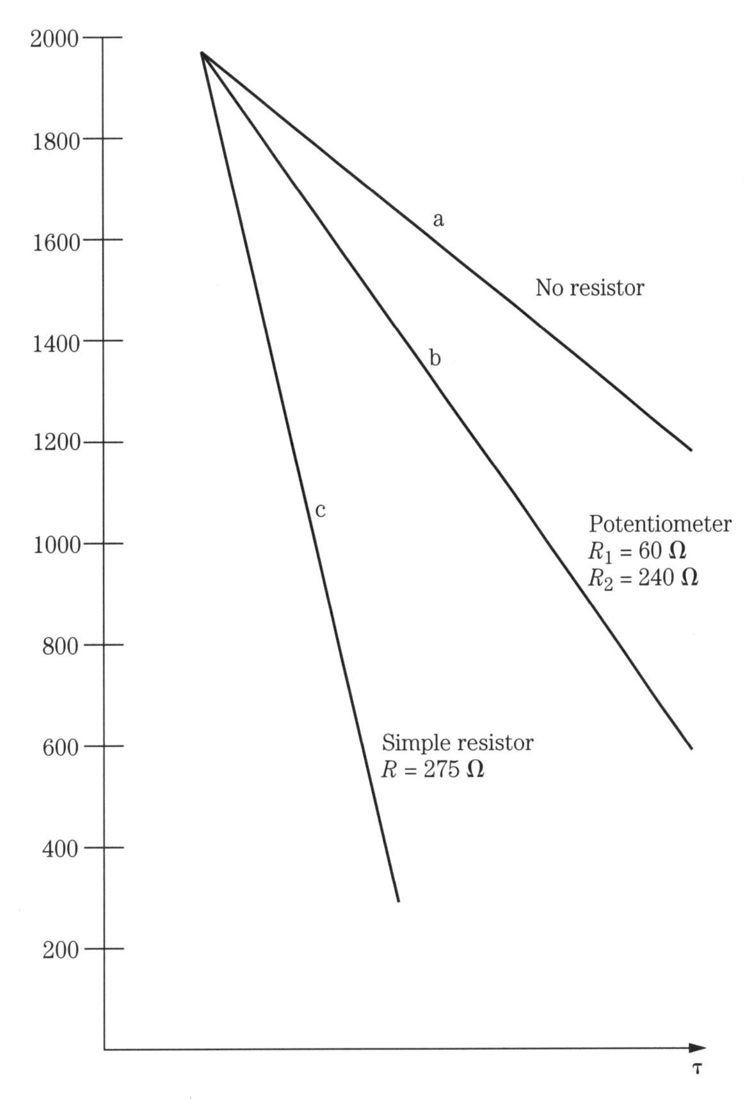

4-2 Speed versus torque curves: (a) V_T held constant, (b) V_T reduced by current flow through potentiometer.

The process of forcing an SCR to turn off is called *commutation*. This is related to, but not the same as, the commutation that is effected by the brushes and commutator of a motor. The SCR will be blocked when the anode voltage is negative, regardless of whether or not the gate voltage is positive with respect to the cathode.

Unijunction transistors

A device that is often used in motor control circuits along with SCRs is a device called a *unijunction transistor* (UJT). As its name implies, a UJT has a single junction as shown in Fig. 4-4. The three terminals are *emitter* (E), *base-1* (B1), and

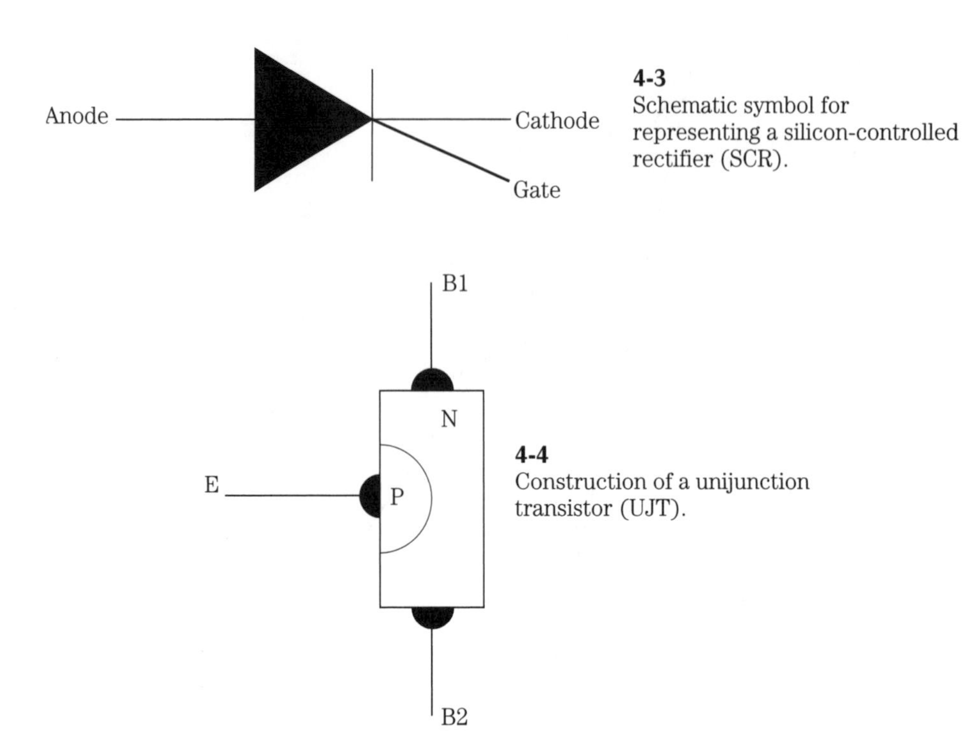

4-3
Schematic symbol for representing a silicon-controlled rectifier (SCR).

4-4
Construction of a unijunction transistor (UJT).

base-2 (B2). The (relatively) large hunk of n-type material acts as a resistor between terminals B1 and B2, with the junction forming a diode. An approximately equivalent circuit for a UJT circuit is shown in Fig. 4-5. The voltage at the node labeled X will be ηV_{BB} where η is the *intrinsic standoff ratio* given by:

$$\eta = \frac{R_{\mathrm{B1}}}{R_{\mathrm{B1}} + R_{\mathrm{B2}}}$$

As long as the voltage at the emitter terminal is less than $V_D + \eta V_{BB}$, the diode will be reverse-biased and no current will flow into the emitter. (The voltage V_D is the *forward drop* of a silicon diode; typically 0.5 to 0.6 V.) Once the emitter voltage is increased to a value greater than $V_D + \eta V_{BB}$, the diode will become forward-biased and begin to conduct. Once the junction becomes forward-biased, holes are injected into the base and effectively decrease R_{B1}. This decrease is dramatic, dropping from an intrinsic value of several thousand ohms to approximately 20 Ω. This behavior of a UJT can be exploited to "square up" a varying voltage applied to the emitter, and in fact the majority of UJTs in motor control circuits are used this way to provide a clean trigger to an SCR. Schematic symbols used to represent UJTs are shown in Fig. 4-6.

Motor control circuitry

An SCR-based circuit for controlling the speed of certain PM motors is shown in Fig. 4-7. This circuit is an improvement over direct rheostatic control of armature voltage because the voltage is regulated via switching rather than via resistive dissi-

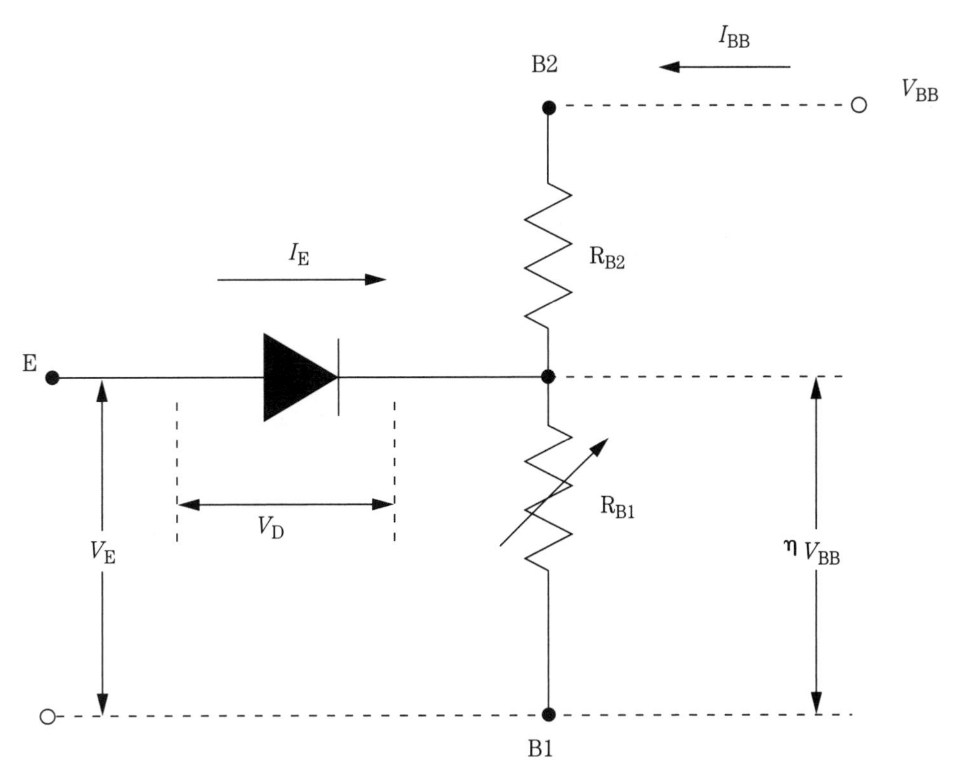

4-5 Equivalent circuit for a unijunction transistor.

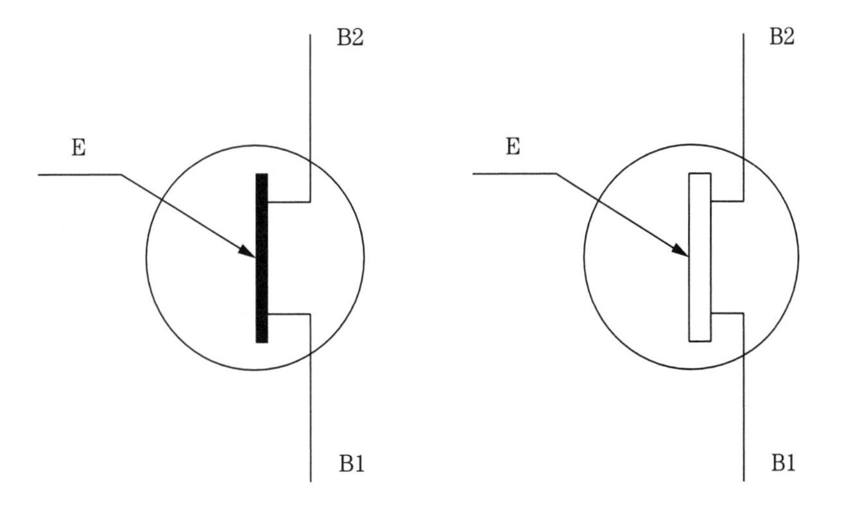

4-6 Schematic symbols used to represent unijunction transistors.

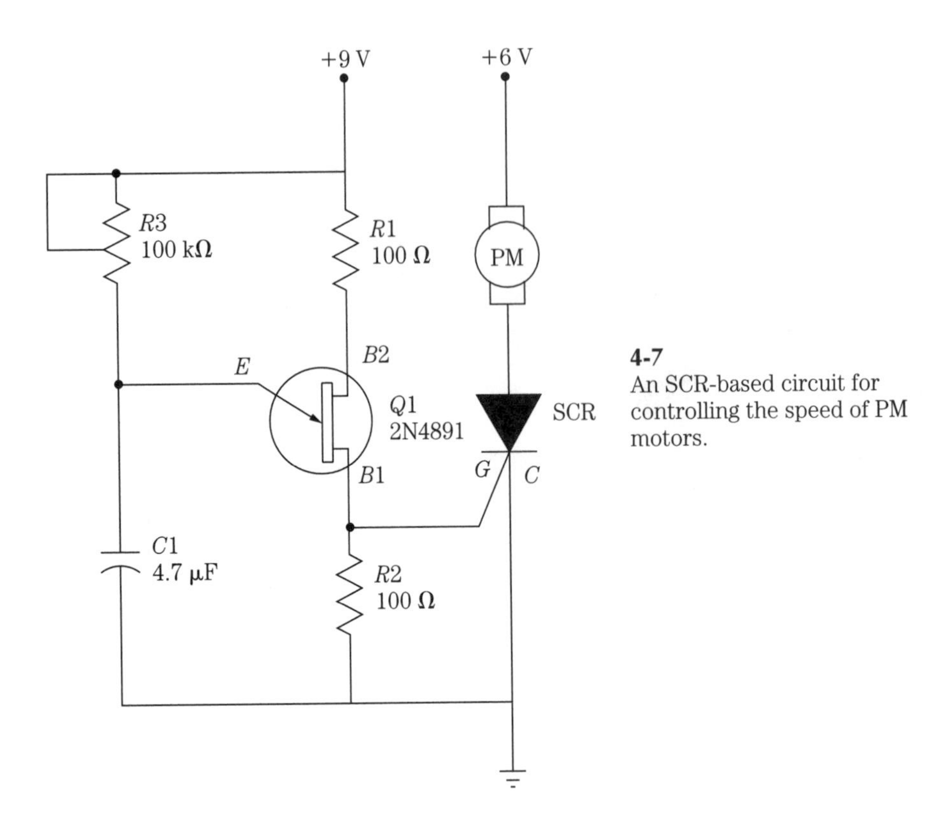

4-7
An SCR-based circuit for controlling the speed of PM motors.

pation. The SCR used should have a current rating (I_{max}) that is larger than the largest armature current that the motor will draw. The UJT together with $R1$, $R3$, and $C1$ forms an oscillator. The capacitor charges through $R3$ until the voltage across the capacitor exceeds the UJT's turn-on voltage of $V_D + \eta V_{BB}$. Because the voltage on the capacitor is the same as the UJT's emitter voltage, this will cause the UJT to fire, conducting current from the emitter to base-1, thereby quickly discharging the capacitor. This will also cause the voltage at $B1$ to rise due to the *IR* drop through resistor $R2$. This increased voltage at $B1$ will cause the SCR to fire, allowing a current to flow through the motor and through the SCR's anode-to-cathode path to reach ground. As the capacitor discharges, the UJT turns off thus allowing the capacitor to begin charging again. Now we get to the explanation of why this circuit works for only certain motors. As configured, the circuit has no provisions for turning off the SCR once it has been triggered. For a great many PM motors, the interruptions to armature current flow caused by the normal gaps between the motor's commutator segments will be sufficient to turn off the SCR. However, for some combinations of motors and SCRs, the motor's commutator interruptions will not be sufficient to cause the SCR to turn off. These motors will run at top speed regardless of how R1 is set.

5
CHAPTER

Stepper motors

Stepper motors or *stepping* motors earn their name from the fact that they are capable of movement in discrete steps. This ability makes them well-suited for use with open-loop control in many different types of positioning applications. There are three main types of stepper motors: *permanent magnet*, *variable reluctance*, and *hybrid*.

Variable reluctance stepper motors

Consider a simple two-pole stator with rotor as shown in Fig. 5-1. When a current is passed through the stator coils, the rotor will be pulled into the position shown in the figure. When the rotor is in this position, the reluctance of the magnetic circuit is minimized.

The basic principle illustrated in Fig. 5-1 is turned into a working *variable reluctance* stepper motor by adding more stator poles and rotor teeth. A simple example having six stator poles and four rotor teeth is shown in Fig. 5-2. Assume the coils on poles **1** and **4** are energized with the rotor in the position shown. To produce a clockwise step, coils **1** and **4** are de-energized and coils **3** and **6** are energized. Because of the shorter distances involved, the attractions between tooth **B** and pole **3** and between tooth **D** and pole **6** are stronger than the attractions between tooth **C** and pole **3** and between tooth **A** and pole **6**. Therefore, the rotor will step 30 degrees clockwise. If instead of energizing coils **3** and **6**, we energize coils **2** and **5**, the rotor will step 30 degrees counterclockwise. Characteristics of variable reluctance stepper motors include:

- No *holding torque* with the stator de-energized.
- Rotation in both directions is possible.
- Single-polarity driving circuitry can be used.
- Low efficiency.
- Low torque-to-size ratio.
- Small step-size designs are possible.
- High slew speeds are possible.

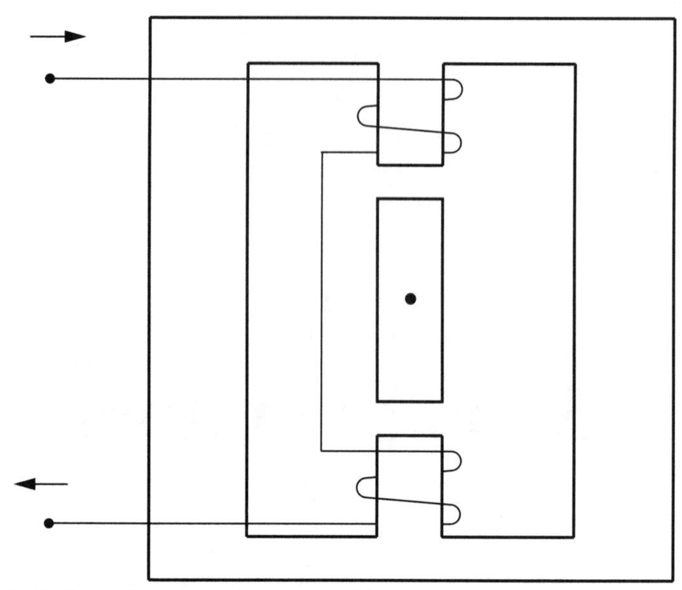

5-1 Two-pole rotor and stator combination in a position that minimizes the reluctance of the magnetic circuit.

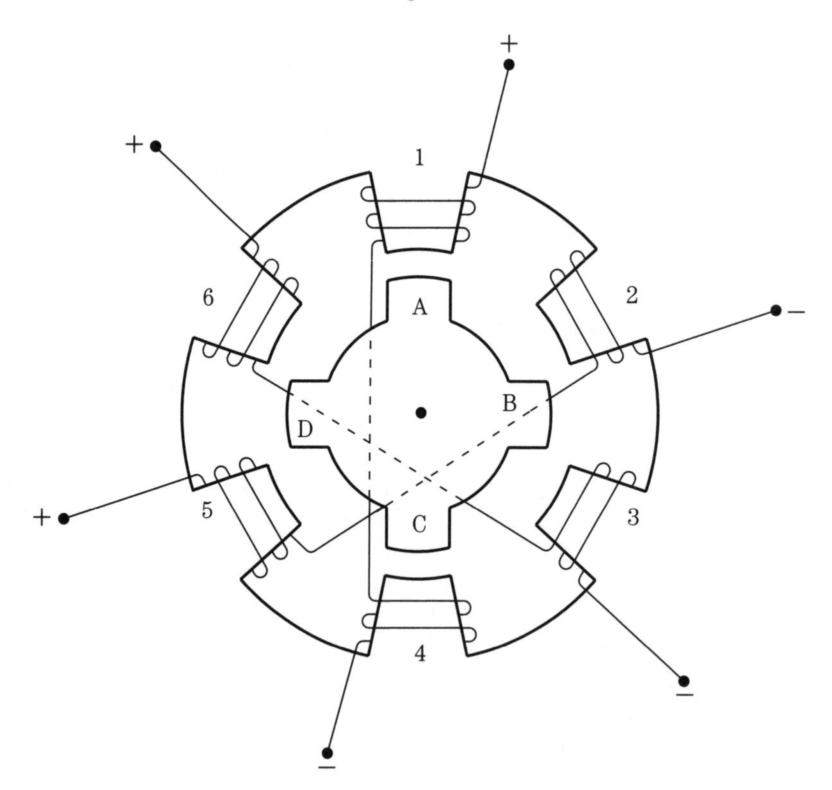

5-2 End view of a simple VR stepper motor having six stator poles and four rotor teeth.

Permanent magnet stepper motors

The structure of a permanent magnet stepper motor is shown in Fig. 5-3. The stator is similar to the stator of a variable reluctance stepper motor. The rotor is a permanent magnet with diametrically opposed poles as shown. Operation is similar to the operation of a variable reluctance motor except for a sensitivity to the polarity of the stator poles. If the motor starts in the position shown in Fig. 5-3, the direction of rotation will depend on the direction of current flow and hence the north/south polarity of the stator poles. If pole **2** is made south, the rotor will step 60 degrees clockwise; if pole **6** is made south, the rotor will step 60 degrees counterclockwise.

Characteristics of permanent magnet stepper motors include:

- Some holding torque with the stator de-energized.
- Rotation in both directions is possible.
- Dual polarity driving circuitry required unless the motor has bifilar stator windings.
- "Good" efficiency.
- High torque-to-size ratio.
- Step sizes must be kept relatively large due to back emf generated by the magnets.

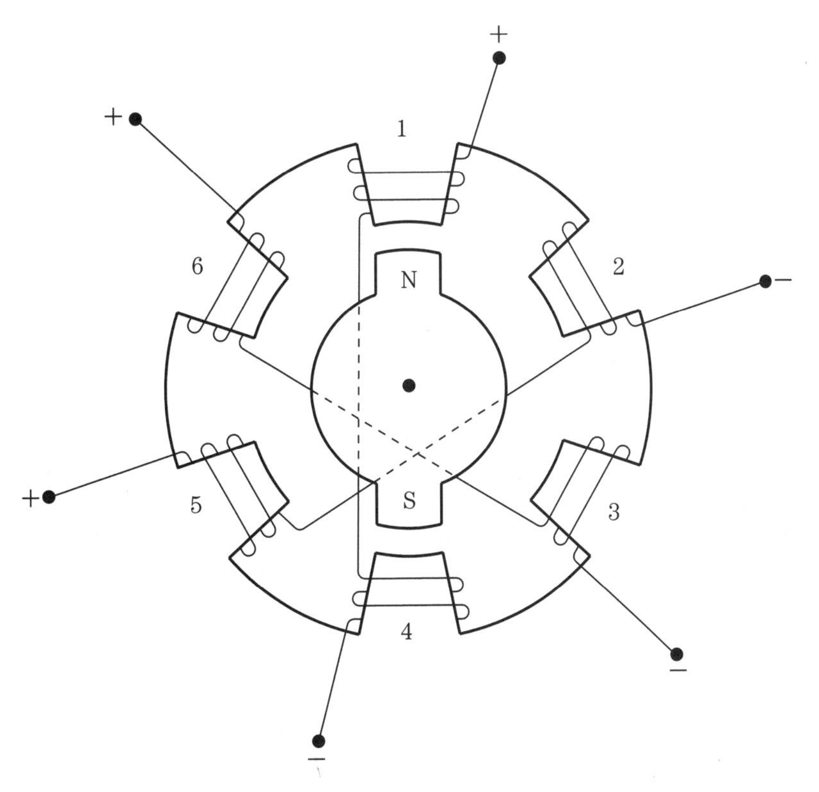

5-3 Internal structure of a permanent stepper motor.

Hybrid stepper motors

In order to understand the principles of operation for hybrid stepper motors, let's examine the internal structure of a simple motor that steps in 30-degree increments. The rotor has two Y-shaped pole pieces like the one shown in Fig. 5-4. These pole pieces are separated by a permanent magnet as shown in Fig. 5-5.

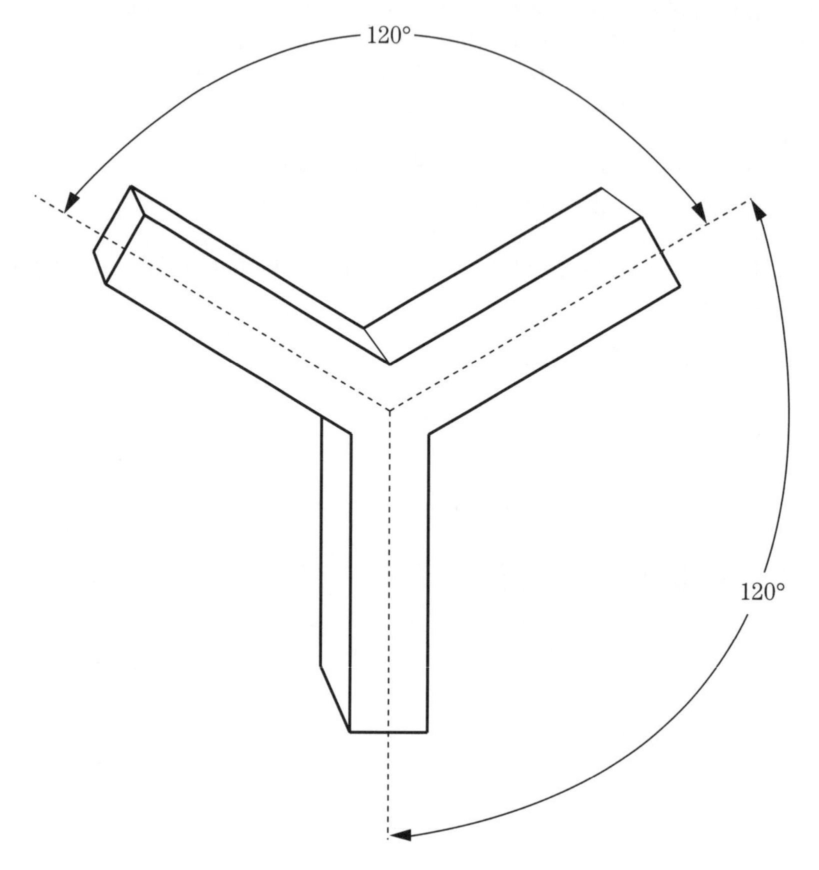

5-4 Pole piece for a 30° hybrid stepper motor.

Locating the pole pieces at the ends of the permanent magnet effectively creates a magnet with a "three-pronged" north pole at one end and a "three-pronged" south pole at the other end. Notice that the two pole pieces are offset so that the poles at one end bisect the angles between the poles of the other end. The stator of the motor consists of a shell having four internal "teeth" or ridges as shown in Fig. 5-6. These ridges run the length of the motor so that the north pole piece of the rotor is close to one end of the ridges and the south pole piece is close to the other end of the ridges.

Coils of wire are wound around these ridges. A single turn of each coil is drawn in Fig. 5-6 to show their relative phasing. The coils on opposite sides of the stator are connected in pairs as shown. The connections are arranged so that a direct current

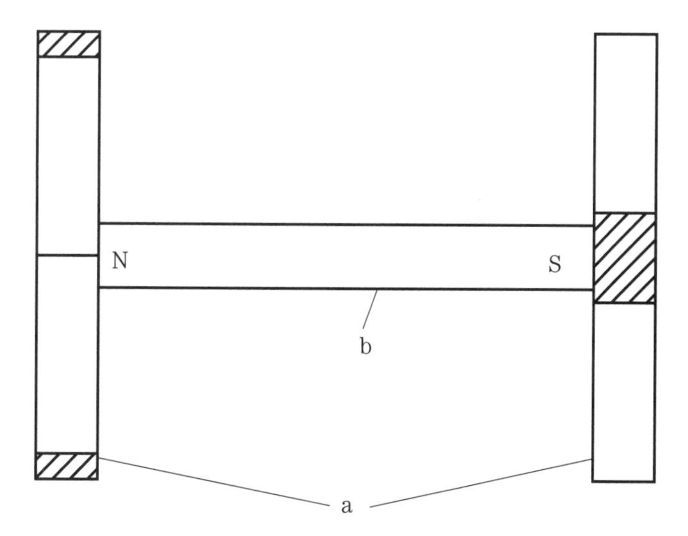

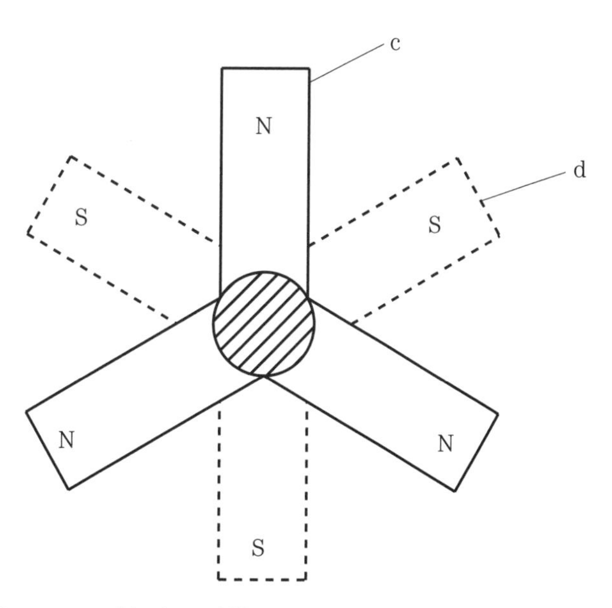

5-5 Rotor assembly for a 30° hybrid stepper motor. Side view shows pole pieces (a) at ends of permanent magnet (b). End view shows North pole piece (c) in foreground offset by 60° from the South pole piece (d) in the background.

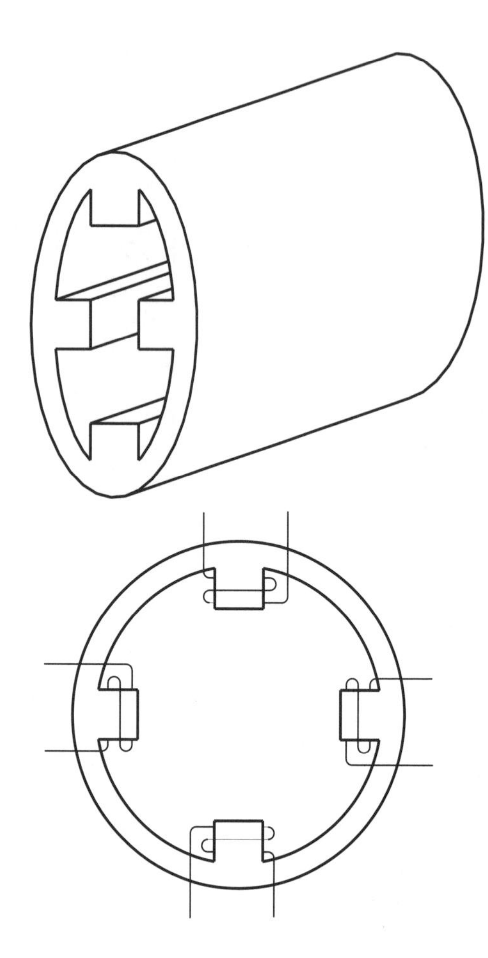

5-6
Stator configuration of a hybrid stepper motor.

through one pair of coils will cause one ridge to become a north pole and one ridge to become a south pole. As depicted in Fig. 5-7, this will cause the rotor to align itself so that one prong of the south pole piece is directly adjacent to the north ridge, and one prong of the north pole piece is directly adjacent to the south ridge.

If the current is shifted to the other pair of coils, the north and south stator poles will shift 90 degrees. This will cause the rotor to move so that once again one prong of the south pole piece is directly adjacent to the north stator pole, and one prong of the north pole piece is directly adjacent to the south stator pole. The direction of rotation is governed by which rotor poles are closest to their new positions that are called for by a shift in stator poles. Let's assume that the stator poles are shifted 90 degrees counterclockwise so that pole **D** becomes north and pole **B** becomes south. Just after the stator field is shifted, the rotor has a "choice" between: (1) rotating counterclockwise 90 degrees to align rotor pole **S1** with stator pole **D**, or (2) rotating clockwise 30 degrees to align rotor pole **S3** with stator pole **D**. The attractive force between **D** and **S3** is greater than the attractive force between **D** and **S1**; therefore the rotor "yields" to the stronger force and rotates clockwise by 30 degrees to the position shown in Fig. 5-8. Thus a 90-degree counterclockwise shift in the stator poles causes a 30-degree clockwise step of the

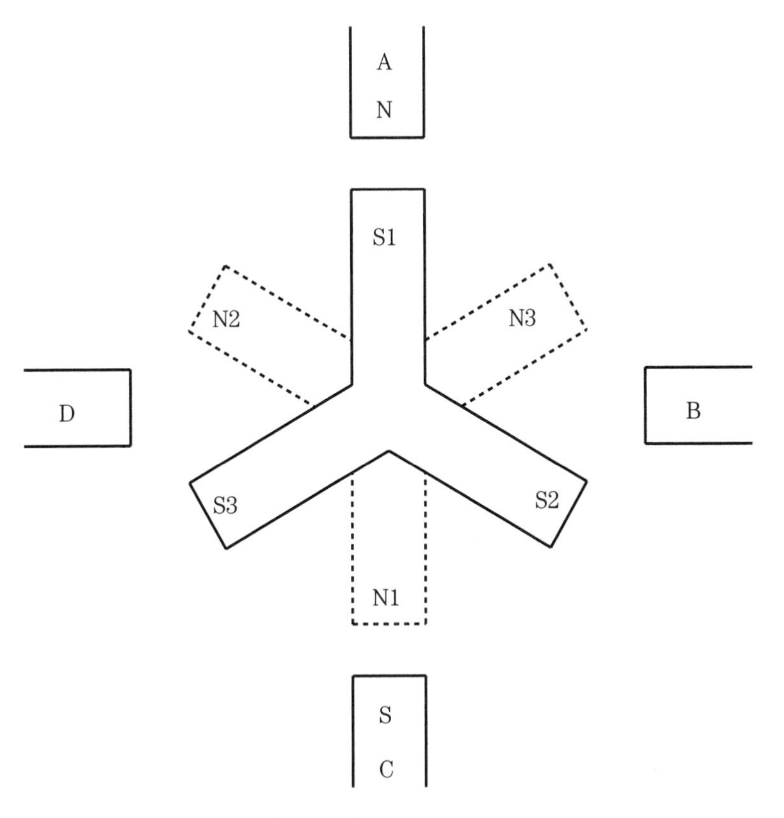

5-7 Rotor position in a 30° hybrid stepper motor.

rotor. Conversely, a 90-degree clockwise shift in the stator poles causes a 30-degree counterclockwise step of the rotor.

It is also possible to operate this motor by simultaneously energizing both pairs of coils. Instead of lining up with rotor poles directly adjacent to a stator pole, the rotor pole will in this case lock up in an intermediate position as shown in Fig. 5-9. In this position, two south rotor poles **S1** and **S2** are equidistant from the two north stator poles **A** and **B**, which seems reasonable given equal attractive forces between **A** and **S1** and between **B** and **S2**. The remaining south rotor pole **S3** lies midway between the south stator poles **C** and **D**, which are both exerting repulsive forces against **S3**. The position of stator poles can be rotated 90 degrees by reversing the polarity of just one pair. For example, reversing the polarity of poles **A** and **C** rotates the stator pattern 90 degrees clockwise, causing the rotor to move one 30-degree step counterclockwise to lock up in the position shown in Fig. 5-10.

Halfstepping

It is possible to *halfstep* a hybrid motor by alternately energizing both pairs of stator windings, then just one pair, then both pairs, etc., etc. to produce a sequence like the one depicted in Fig. 5-11.

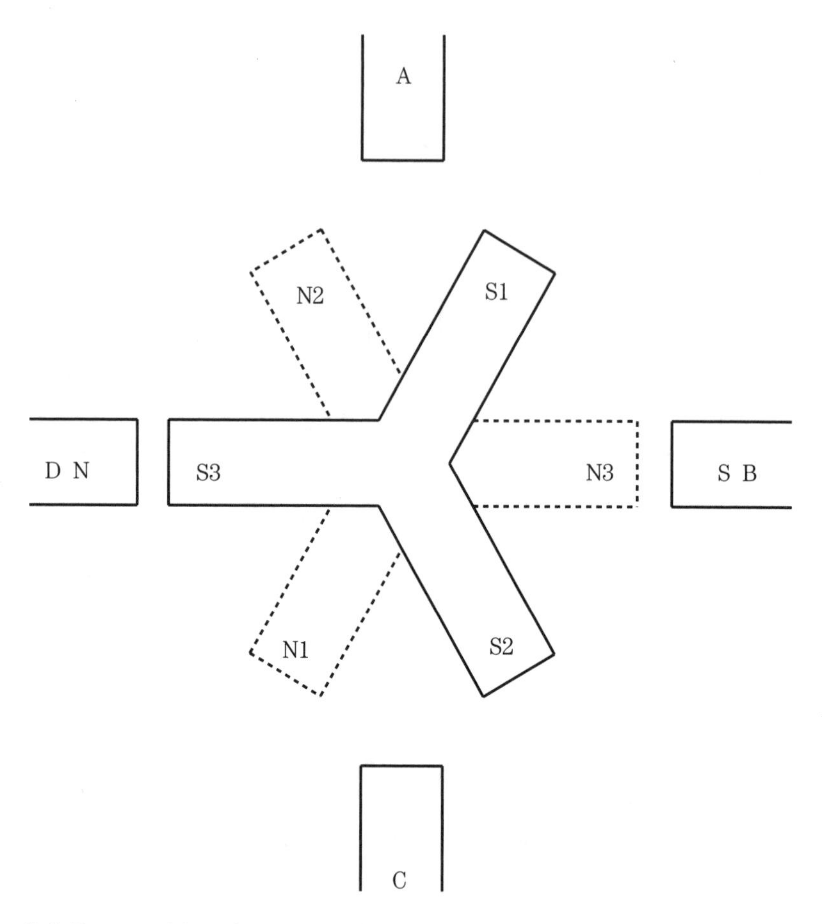

5-8 Rotor position after stator poles are shifted 90° counterclockwise.

Characteristics

Characteristics of a hybrid stepper motor include:

- Moderate holding torque with the stator de-energized.
- Rotation in both directions is possible.
- Dual polarity driving circuitry required unless bifilar stator windings are used.
- "Good" efficiency.
- High torque to size ratio.
- Small step sizes are possible.
- High slew speeds are possible.

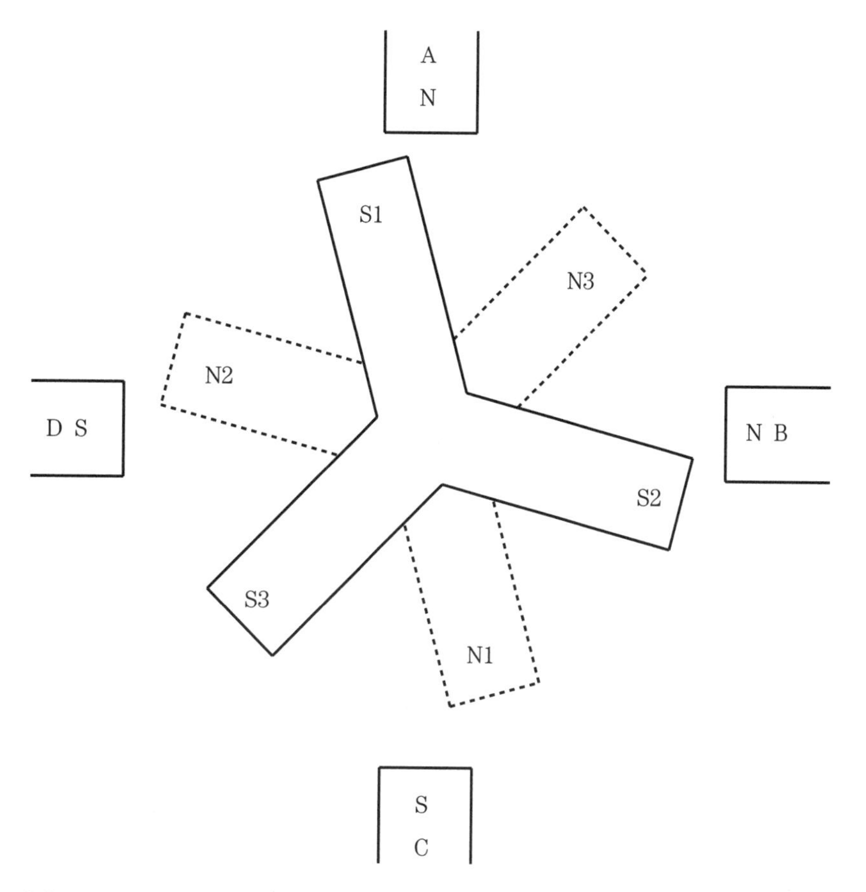

5-9 Rotor positioning in a hybrid stepper motor with both pairs of stator poles energized.

Stepper motor drives

In order to take advantage of the positioning capabilities of the various types of stepper motor, it is essential to match the control and drive circuitry to the type of motor being used.

Control logic

At the logic level there are several different ways to architect a stepper controller. Two possibilities are depicted in Fig. 5-12. Controller **A** accepts two logical inputs. The line labeled as "CW/$\overline{\text{CCW}}$" determines the direction of stepping. A **true**

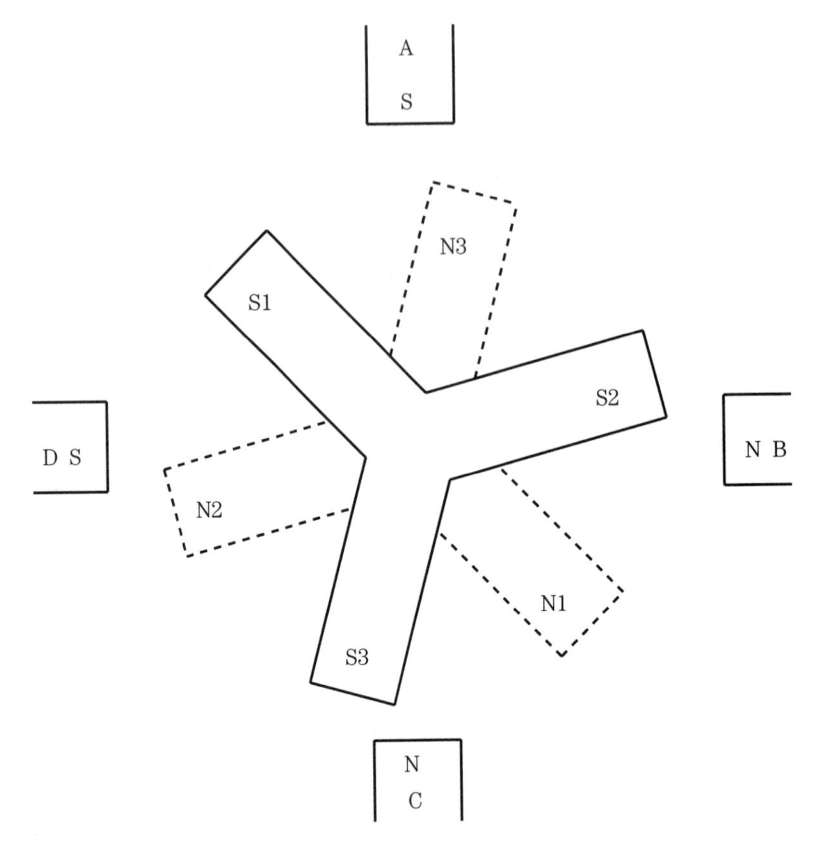

5-10 Rotor positioning after stator pole pattern is rotated 90° CW by reversing the polarity of poles **A** and **C**.

input indicates that clockwise rotation is desired, and an input of **false** indicates that counterclockwise rotation is desired. The line labeled "step" must be pulsed once for each desired step. For frequent multistep rotations, this approach can become a burden for the central controller that provides the step pulses.

Controller **B** has an input for the desired number of steps, so multistep rotation can be set up and launched with a single transfer from the central controller. The outputs of either controller type depend upon the specific type of stepper motor to be controlled. As noted in the discussions of the various motor types, most stepper motors have a two-phase stator. However, two-phase motors can come in 4-lead, 5-lead, 6-lead, and 8-lead versions. The internal connections for each configuration are shown in Fig. 5-13.

For bidirectional operation of a 4-lead motor, it is necessary to use a bipolar drive. Voltages of one polarity are applied to the coils for rotation in one direction, and this polarity is reversed to cause rotation in the opposite direction. Bidirectional operation of 5-, 6-, and 8-lead motors can be achieved using a unipolar drive. For the 5-, 6-, and 8-lead motors shown in Fig. 5-13, voltage is applied to coils **1A** and **2A** for rotation in one direction and to coils **1B** and **2B** for rotation in the other direction. The drive voltage maintains a constant polarity, but an effective reversal is achieved by having the **A** and **B** coils connected in opposite directions.

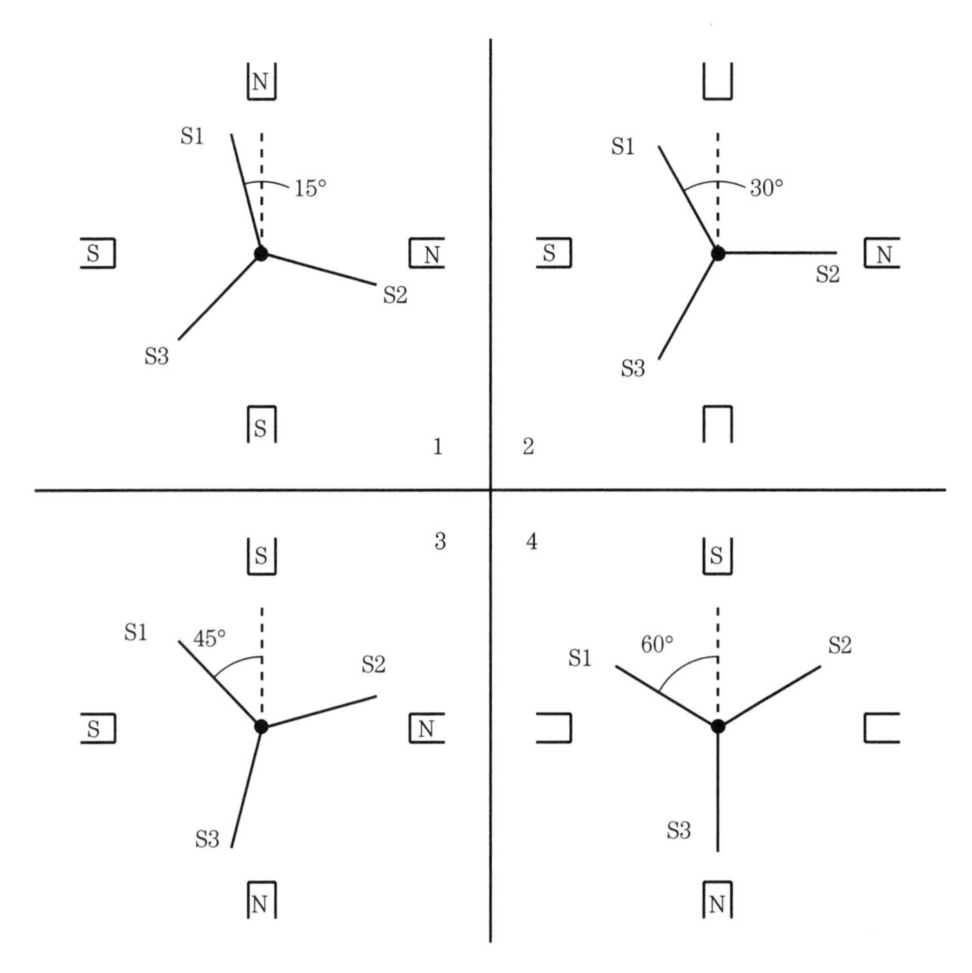

5-11 Sequence of pole positions for half-stepping a 30° hybrid stepper motor.

Drive power

Rarely, if ever, will the outputs of a computer or other digital logic device be directly usable for driving a stepper motor. The logic level must be used to control some sort of electronic switching element that can handle the voltages and currents needed by the motor. A simple transistor switch is shown in Fig. 5-14. An npn transistor is used in a common-emitter circuit.

The coil shown represents one of the stator windings in a stepper motor. When the control input **D** is held low, the transistor is in cutoff and no current flows in the coil. If a positive-going logic pulse is applied to the control input **D**, the base-emitter junction will become forward-biased for the duration of the pulse, and a pulse of current will flow in the coil. Four of these transistor switches can be used to control 5-, 6-, or 8-lead bifilar-wound stepper motors. The **D** inputs can be driven by TTL logic that controls the sequencing of pulses to the various windings. This scheme works well for low-speed operation, but for high-speed operation, something a bit more complicated may be needed.

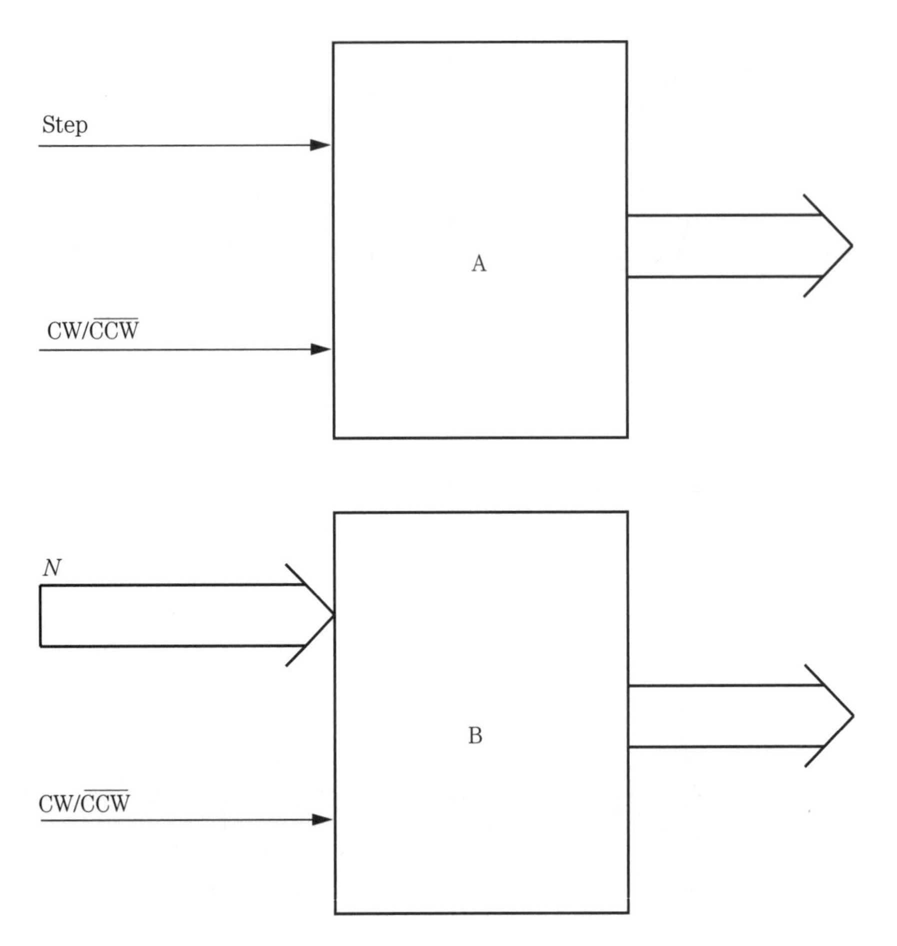

5-12 Stepper motor controllers.

Each of the stator windings has a certain amount of inductance which impedes an instantaneous change in the current flow. If we assume that the transistor switching times are very fast, the current waveform through one of the coils will be as shown in Fig. 5-15. For low-speed operation with fairly long pulses, the nonzero rise and fall times will not cause a problem. However, for high-speed operation, we need narrow pulses as shown in Fig. 5-16. In this case, because of the inductance, the current never really gets all the way up to full scale before the logic pulse ends and the transistor switch turns off. Operation with these stunted pulses will severely limit the torque available from the motor.

Using surplus motors

Often surplus dealers are one of the best sources for affordable stepper motors. Unfortunately, surplus steppers often come without any documentation to tell which leads are which or how the motor should be connected to the driving circuits. Referring back to Fig. 5-13, we can see that simple resistance measurements could be used

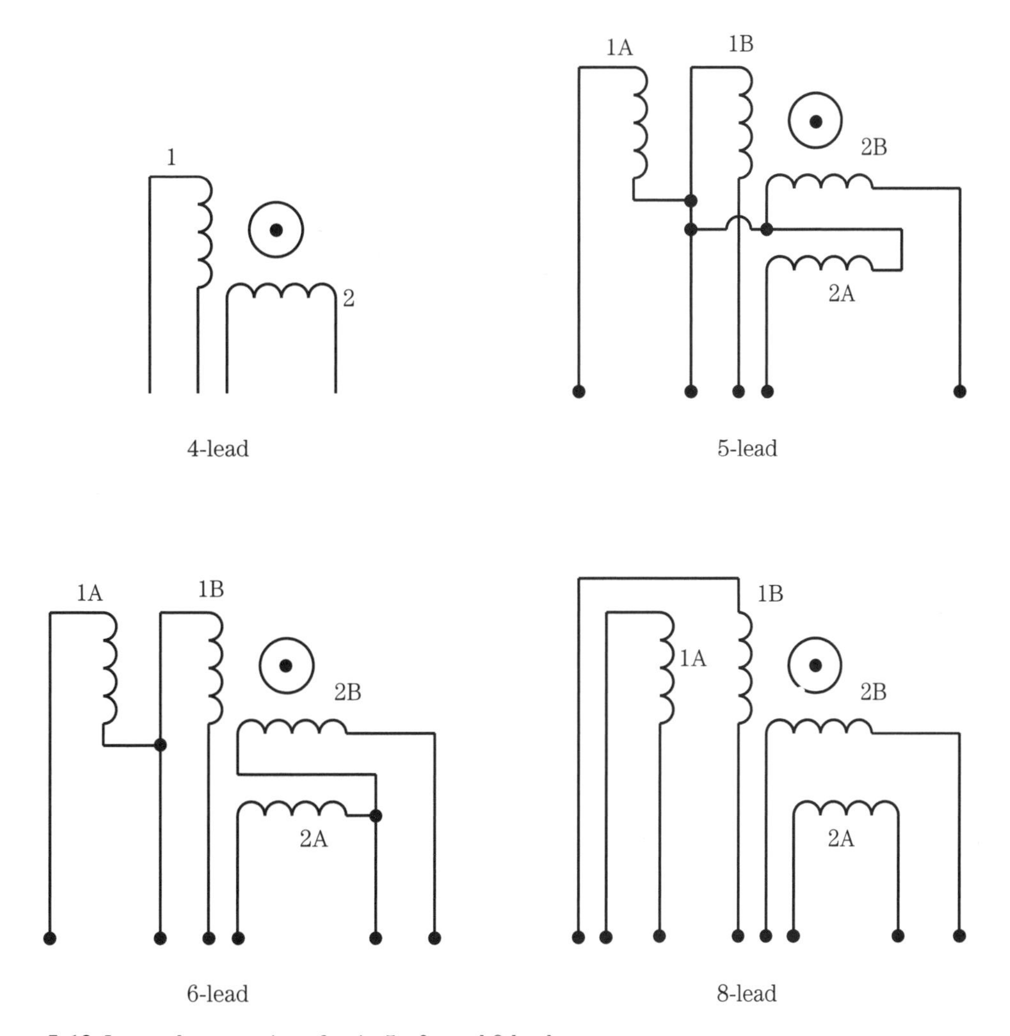

5-13 Internal connections for 4-, 5-, 6-, and 8-lead stepper motors.

to determine which leads are connected to the ends of each coil. For leads that are connected to opposite ends of the same coil, the ohmmeter will indicate the resistance of the coil. For 4-lead and 8-lead motors, an open circuit will be indicated for measurements between leads connected to different coils. For 5-lead and 6-lead motors, some pairs of leads will be connected via a series combination of two coils. The resistance reading across these leads should be about twice the resistance measured across a single coil. It is a simple matter to sort out the two finite resistance values and, by referring to Figs. 5-17 and 5-18, determine how the leads are connected to the coils.

Another simple test will determine whether or not an undocumented motor is of the variable reluctance type. Using your fingers, gently turn the shaft of the motor. If the shaft turns smoothly without any detents or bumps, the motor is probably a variable reluctance design. On the other hand, if detents are felt, the motor is either a permanent magnet or hybrid design.

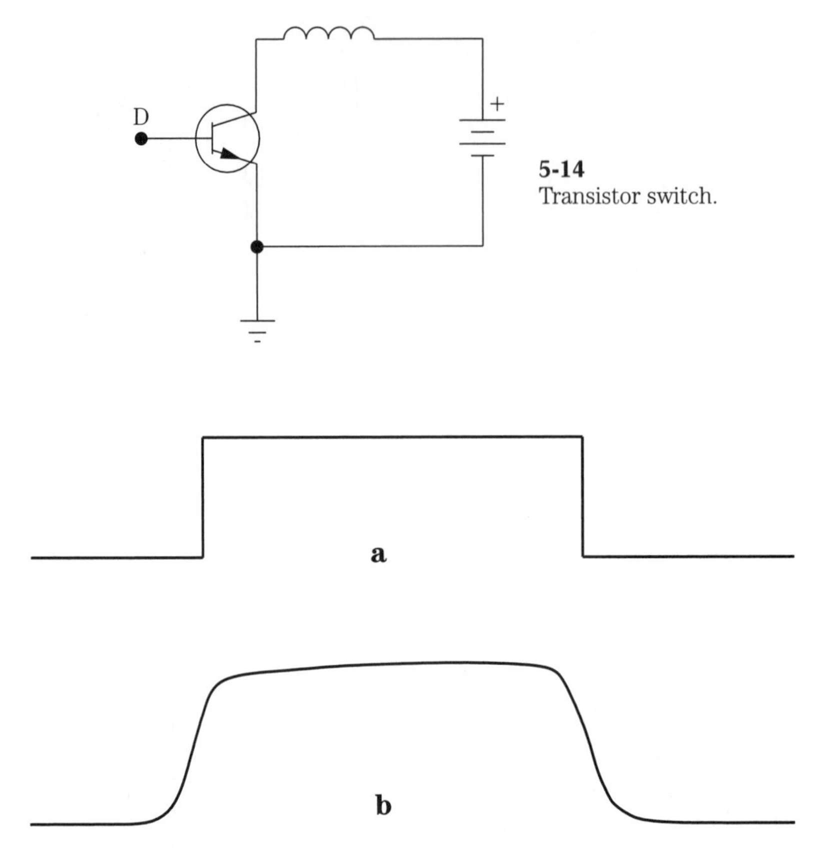

5-14
Transistor switch.

5-15 Stepper control signals when switching is fast relative to pulse duration: (a) control input, (b) output to motor.

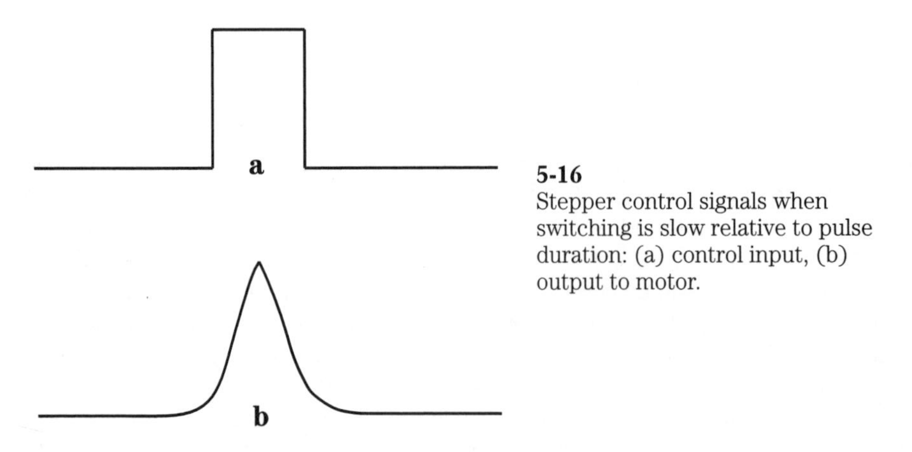

5-16
Stepper control signals when switching is slow relative to pulse duration: (a) control input, (b) output to motor.

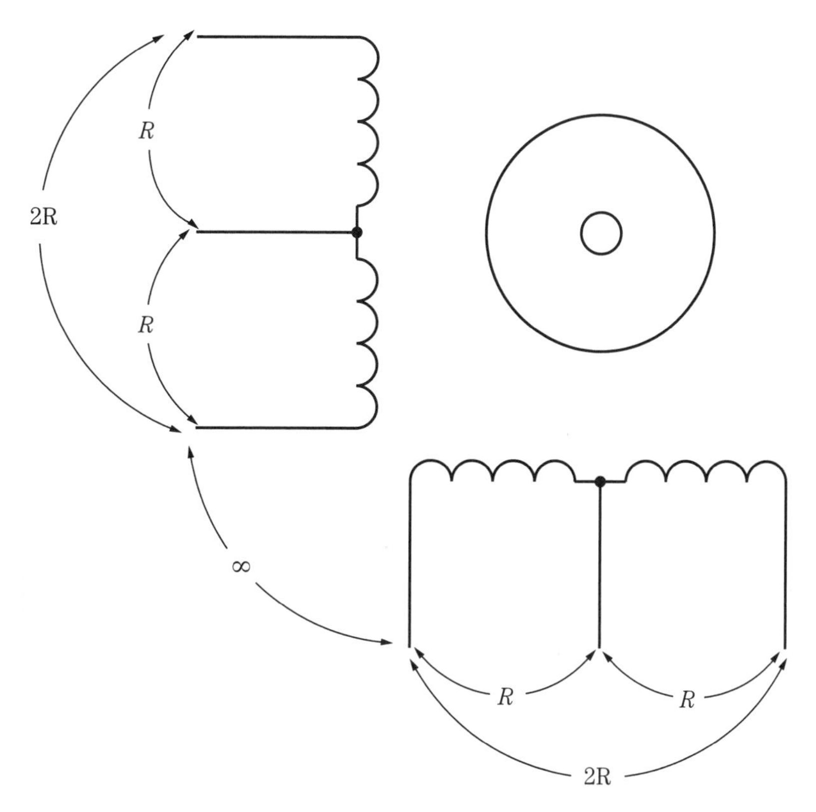

5-17 Resistance combinations for a 6-lead stepper motor.

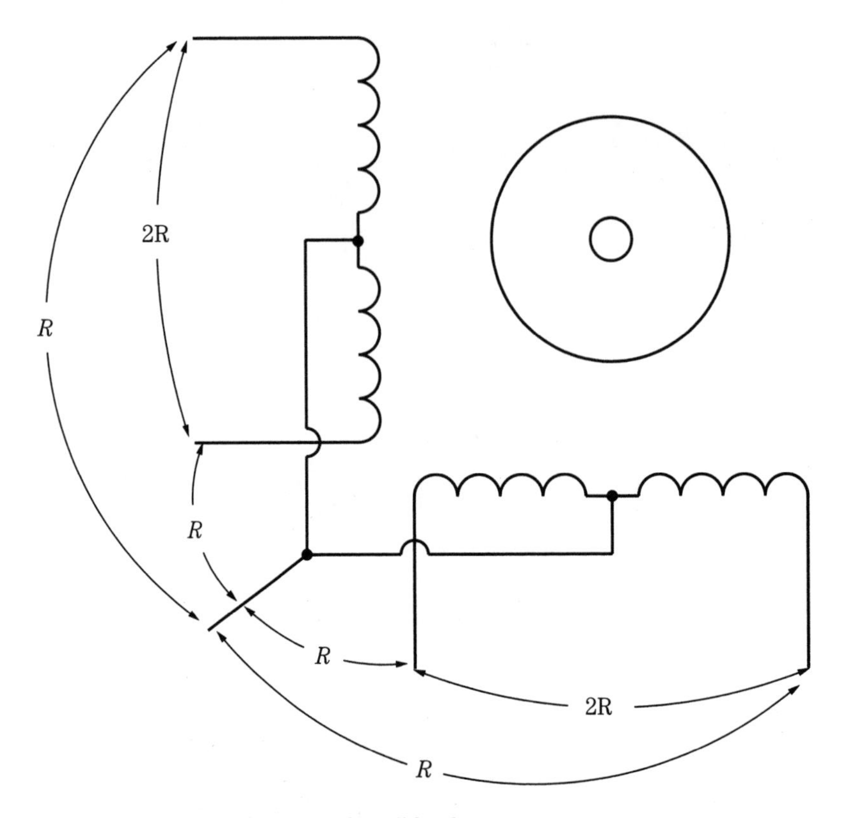

5-18 Resistance combinations for a 5-lead stepper motor.

6
CHAPTER
Solenoids

Strictly speaking, a *solenoid* is any helical winding of wire. Within the context of electromechanical devices, the term is more often used to identify a coil and plunger arrangement such as the one shown in Fig. 6-1. When the coil is energized, the plunger will be pulled toward the center of the coil. If everything is designed just right, there will be a significant amount of force available for this pull, and loads attached to the plunger can be moved along with the plunger. Often a spring will be used to return the plunger to its original position when the coil is de-energized. The distance moved by the plunger is called the *stroke* of the solenoid.

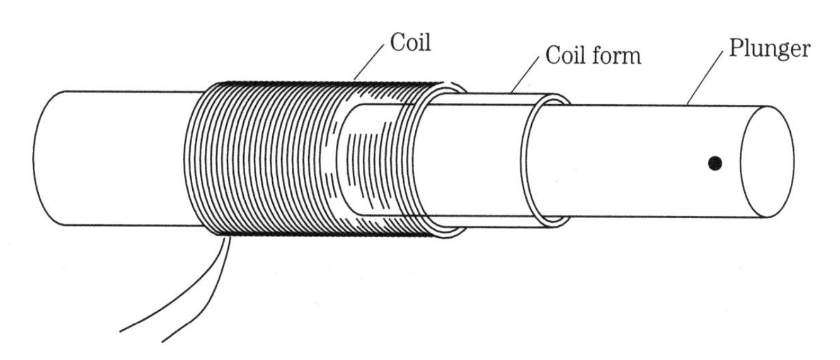

6-1 A simple solenoid.

Homemade solenoids

The precise design of a solenoid can be a very complicated problem, even for professional engineers. However, the design and construction of homebrew solenoids is ideally served by the methodology of "design a little, build a little, test a little, and tweak as necessary."

Design parameters

There are quite a few design parameters associated with a simple solenoid. Some of these include:

- number of turns in the coil
- gauge of the wire used in the coil
- dimensions of the coil
- dimensions of the plunger
- stroke length
- available force
- magnetic properties of the shell and plunger
- voltage across the coil
- current through the coil

As we might expect, many of these parameters are interrelated and an improvement in one parameter may come at the cost of a degradation in another parameter. The homebrew design process is not completely random; there are some guidelines and practical rules-of-thumb.

Coil turns

The force exerted by either a solenoid or an iron core electromagnet is approximately proportional to the square of the number of turns:

$$F \propto N^2$$

In other words, doubling the number of turns in the coil will approximately quadruple the available force. This of course assumes that the current remains constant. However, doubling the number of turns will approximately double the resistance of the coil and therefore halve the current, providing that nothing else is changed. The current can be held constant by either doubling the applied voltage or using a larger gauge wire to halve the resistance of the doubled coil. Everything comes at a price; increasing the voltage may lead to excessive heating in the coil.

Wire gauge

Small diameter wire allows more turns to be placed on a given core, but smaller diameter wire will have a higher resistance than larger diameter wire. Any homebrew design should definitely use enamel-insulated magnet wire of some gauge, rather than plastic-insulated bell wire or hookup wire. The thickness of the plastic insulation will make it difficult to get lots of turns packed into a small volume around the core.

Current

The force exerted by either a solenoid or an iron core electromagnet is approximately proportional to the square of the current through the coil:

$$F \propto I^2$$

In other words, doubling the current will approximately quadruple the available force.

Physical dimensions

The stroke length of a solenoid depends upon the dimensions of the coil and plunger. Generally, longer plungers are capable of longer strokes. (When speaking of plunger length, we are only considering the length of the cylinder of magnetic material. The linkage that connects the plunger to the load is nonmagnetic and is not part of the plunger.) Consider the following:

- Once the plunger is completely inside the coil, the magnetic forces acting on the plunger are too weak to cause further movement of the plunger deeper into the coil. Therefore, it makes no sense to have a coil that is longer than the plunger.
- If the plunger is longer than the coil, the forces will tend to center the plunger in the coil (i.e., leave approximately equal lengths of plunger sticking out of both ends of the coil). In many designs, stops are included to prevent the leading end of the plunger from going past the end of the coil. (The leading end is the end that lies just inside the coil when de-energized, and which is the "forward" end relative to the direction of motion as the plunger moves in response to the coil becoming energized.) Just how "centered" the plunger will be depends upon the weight of the plunger and the force of the load attached to the trailing end of the plunger. Theoretically, an unloaded plunger with zero mass should center exactly. However, plungers have nonzero mass. Even when unloaded, plungers that are several times longer than their coils tend to stop moving a good bit before center.
- In the de-energized "at rest" position a part of the plunger must lie inside the coil. If the plunger is completely outside the coil, the coil will not generate sufficient magnetic force to pull in the plunger.

Figure 6-2 shows the relationships between the various dimensions and plunger positions. For the "ideal" limiting case, the stroke length L_S would be given by:

$$L_S = \frac{L_C + L_P}{2} \qquad (6\text{-}1)$$

where

L_C = coil length
L_P = plunger length

In all practical cases, the stroke length will always be shorter than the value given by Eq. 6-1. The sensible way to approach this would be to analyze the intended application to determine the **required** stroke length L_R, and then select the coil length L_C and plunger length L_P so that they satisfy:

$$\frac{L_C + L_P}{2} > K_1 L_R$$

and

$$L_P > K_2 L_C$$

where K_1 and K_2 are "fudge factors" that vary between 1.2 and 1.5. For smaller solenoids, values near 1.2 are fine. For larger solenoids, values closer to 1.5 should be

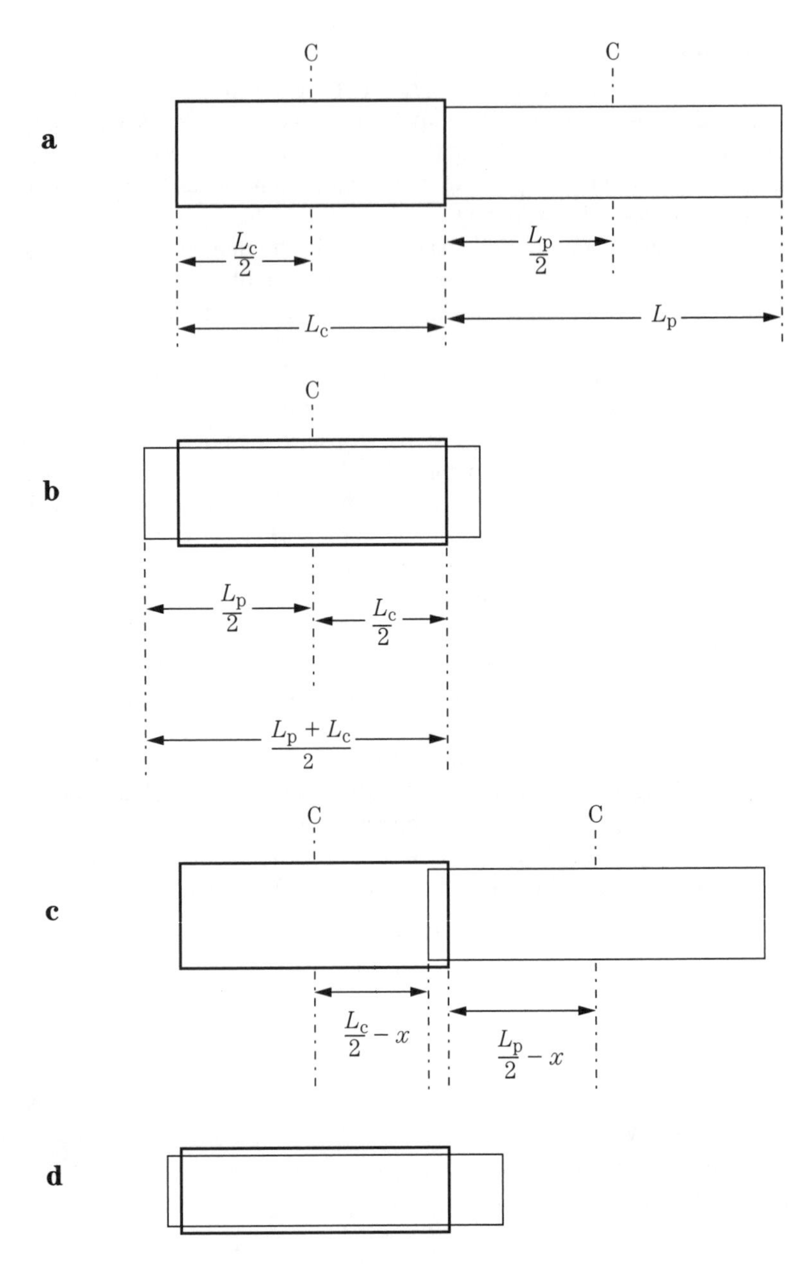

6-2 Plunger positions: (a) "ideal" de-energized position, (b) "ideal" energized position, (c) practical de-energized position, (d) practical energized position.

used. If in doubt, use a larger value because this will result in a longer realized stroke length. If the stroke length winds up being too long, the motion will just hit the stops a little bit harder (there should **always** be stops). If the stroke length winds up being too short, you will have to start over and build a new solenoid.

Shells

For every homebrew solenoid that I have ever made, the coils have been wound on some sort of nonmetallic core form—usually a length of small diameter plastic tube. The plungers have been pieces of steel rod stock. The most powerful commercially produced solenoids have their coils enclosed in a ferrous alloy shell which provides a low reluctance path for the coil's magnetic flux. There is a compromise design that lies between a shell-less homebrew solenoid and a commercial solenoid with a full shell. This design is called an *open frame* solenoid. An open frame solenoid will produce more force than a comparably sized shell-less solenoid and less force than a comparably sized solenoid with a full shell. There are two major variations of the basic open frame design: the C frame style shown in Fig. 6-3 and the D frame style shown in Fig. 6-4. Commercial designs achieve slightly higher forces with the D style, but for homebrew designs, the extra performance is probably not worth the increased complexity. Homebrew solenoid frames can be easily fabricated from the steel angle brackets sold in hardware stores.

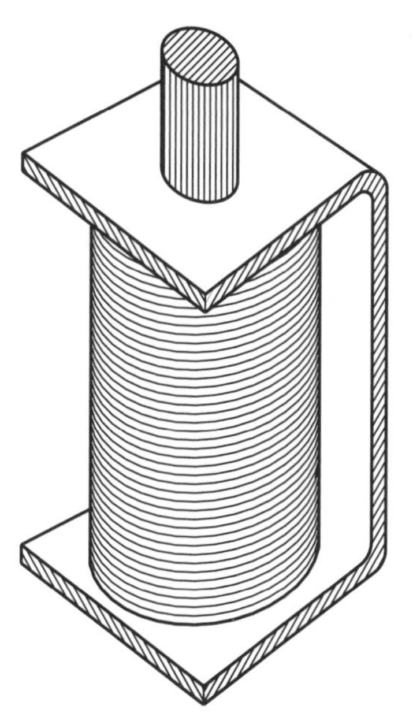

6-3
An open frame solenoid with a C style frame.

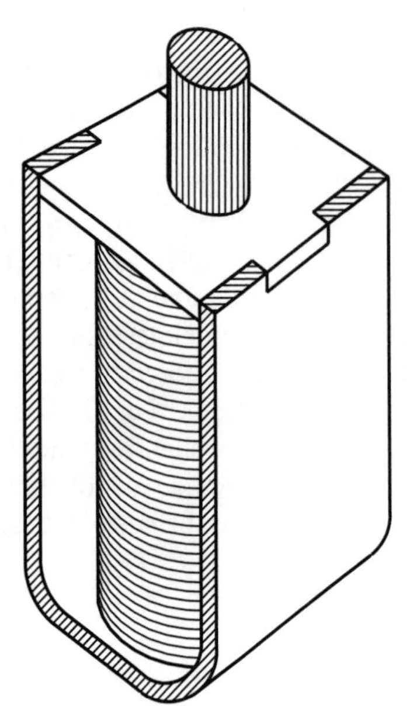

6-4
An open frame solenoid with a D style frame.

7
CHAPTER
Gears and pulleys

Friction drives

Two discs can be used as shown in Fig. 7-1 to transmit rotary motion between two shafts. Let's assume that the discs, or at least their edges, are made of rubber or some similar material that will ensure good nonslip frictional contact between the discs. The speed at which the circumference of one disc moves past a fixed point is simply the length of the circumference times the rotational rate of the disc:

$$s = cf = \pi df$$

where

c = circumference of the disc
f = rotational rate of the disc
d = diameter of the disc

Assuming that the contacting discs cannot slip, their circumferences must move through the point of contact at the same rate. Therefore, we can write:

$$c_1 f_1 = \pi d_1 f_1 = c_2 f_2 = \pi d_2 f_2$$

$$d_1 f_1 = d_2 f_2$$

$$\frac{d_1}{d_2} = \frac{f_2}{f_1}$$

The ratio of rotational rates is inversely proportional to the ratio of the disc diameters. Neglecting frictional losses, the ratio of torques exerted by the discs will be proportional to the ratio of their diameters:

$$\frac{\tau_1}{\tau_2} = \frac{d_1}{d_2}$$

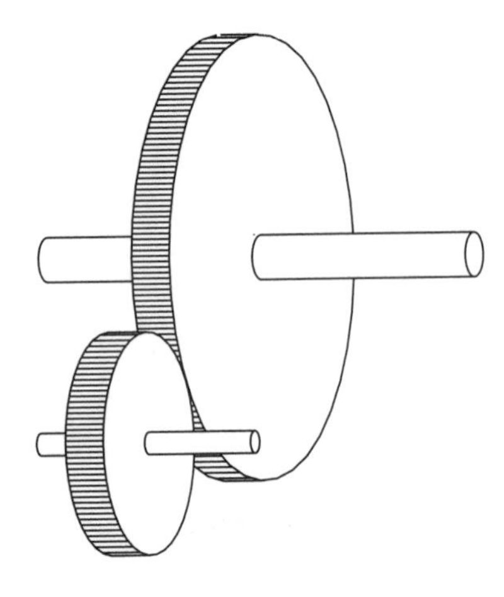

7-1
Two discs used to transmit rotary motion between two shafts.

Example 7-1

Consider the case of a ½-inch-diameter rubber wheel that is driven at 3600 rpm by a motor. This wheel in turn drives a 3-inch-diameter wheel. What is the rotational speed of the 3-inch wheel? If the load on the 3-inch disc exerts a torque of 10 in•oz, what torque must the motor supply to the ½-inch disc?

Solution:

$$d_1 = 0.5,\ d_2 = 3.0, f_1 = 3600,\ \tau_2 = 10$$

$$f_2 = \frac{d_1 f_1}{d_2} = \frac{(0.5)(3600)}{3} = 600 \text{ rpm}$$

$$\tau_1 = \frac{d_1 \tau_2}{d_2} = \frac{(0.5)(10)}{3} = 1.67 \text{ in}\cdot\text{oz}$$

Frictional drives are not limited to use with parallel shafts. For perpendicular shafts, a small wheel can turn against the face of a large wheel as shown in Fig. 7-2a. Conical drive wheels like those shown in Fig. 7-2b can be made for operation at virtually any desired shaft angle.

As shown in Fig. 7-3, the basic wheel drive concept can be extended to gear drives by adding teeth to the wheels. A belt drive is obtained by separating the wheels and connecting them with a loop of flexible belting. When both the belt and tooth features are added, the result is a chain and sprocket drive. Each of these types of drives will be discussed in subsequent sections.

Sources

There are lots of good sources for wheels that can be used as drive components: wagons, baby strollers, and construction toys to name a few. Wheels can be fabricated in custom sizes out of wood or plastic with a strip of rubber glued around the edge. Small rubber wheels called *capstans* can be found in cassette players and VCRs.

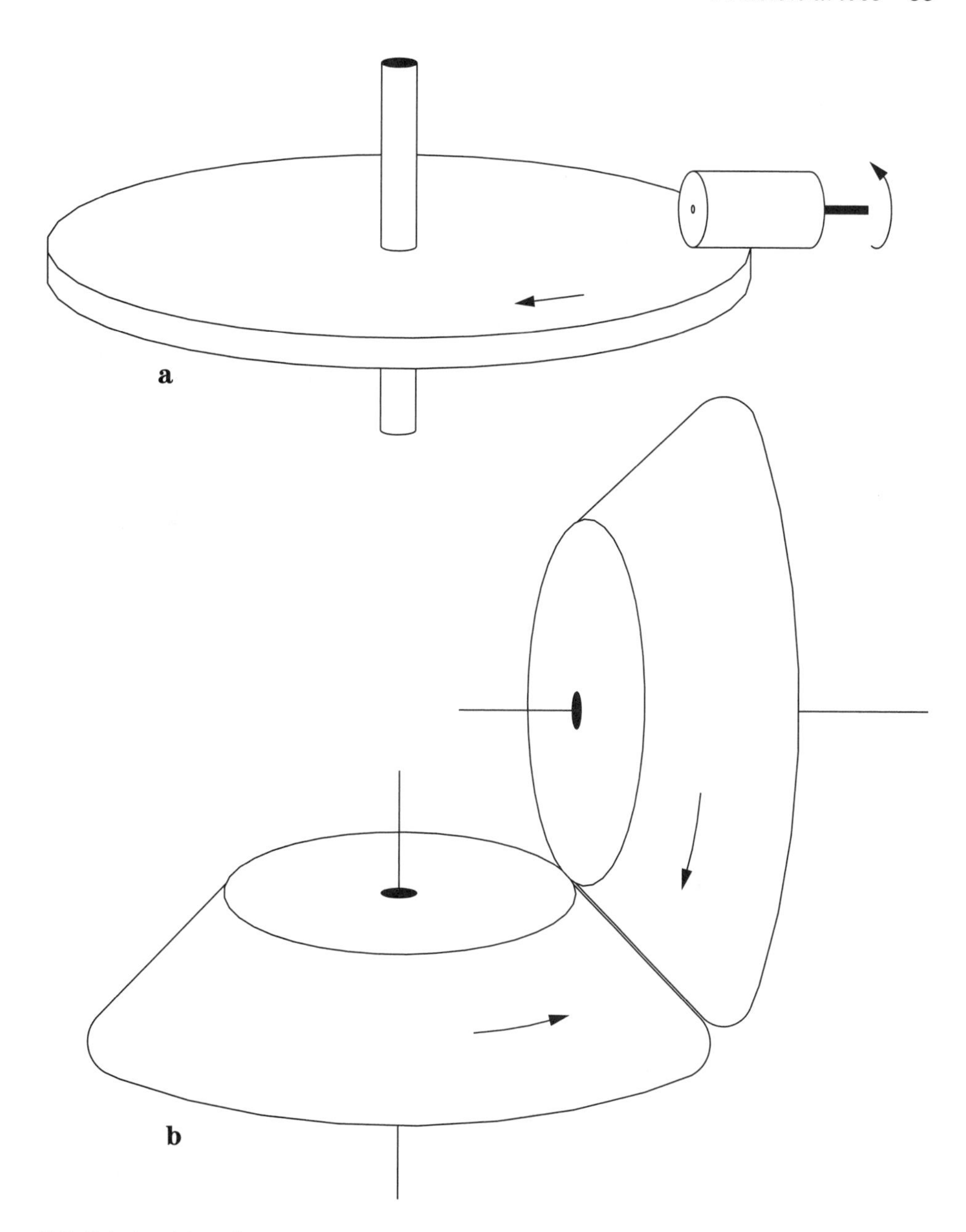

7-2 Friction drives for coupling perpendicular shafts.

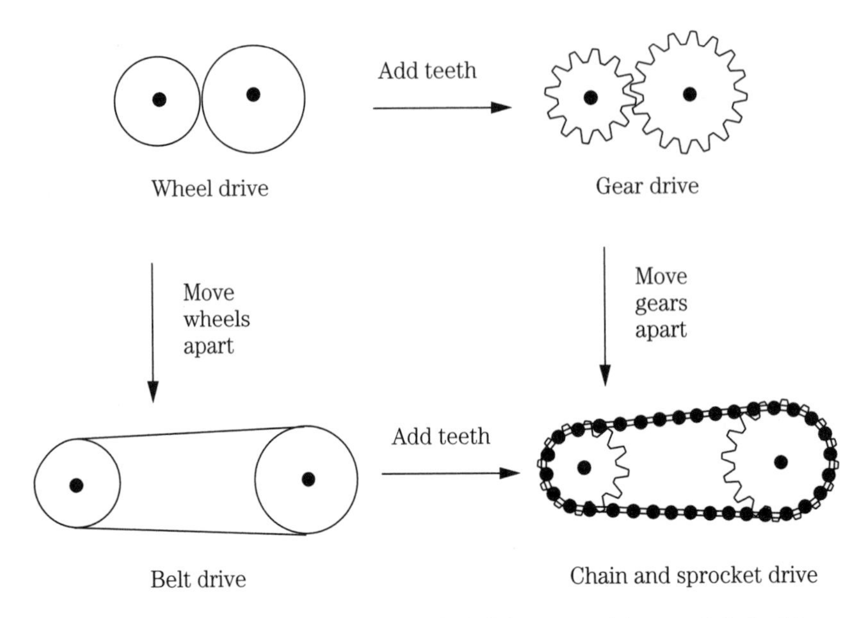

7-3 Relationships between wheel drives, belt drives, gear drives, and chain drives.

Belt drives

Belts and pulleys can be used as shown in Fig. 7-4 to transmit rotary motion from one shaft to another. Belt drives found around the house range from tiny little affairs with belts that are not much more than glorified rubber bands used in cassette players, to the heavy-duty belt drives used in cars to run the alternator, air conditioner, and water pump. For the amateur experimenter, it is fairly easy to construct a belt drive of any required size. Unlike gears, which must be purchased or scavenged, the pulleys for a belt drive can often be fabricated in the home shop. More on this later; first, let's look at the mechanical properties important in design and operation of belt drives.

Round belts

Round belts are the most versatile type of belt for use in the low power, low speed applications that are typically of interest to the amateur experimenter. As their name implies, round belts have a round cross section. Round belts usually do not require idlers or other special devices to maintain tension, and their round cross section is a distinct advantage for serpentine drives, reverse bends, or pulleys on nonparallel shafts. Round belts are usually made from Neoprene, Buna Nitrile, ethylene-propylene, or polyurethane. A serviceable round belt can often be made of thick-walled rubber or plastic tubing that is joined into an endless loop with a suitable connector. In this case, the minimum diameter of the pulleys must be large enough to accommodate the region of reduced flexibility created by the connector.

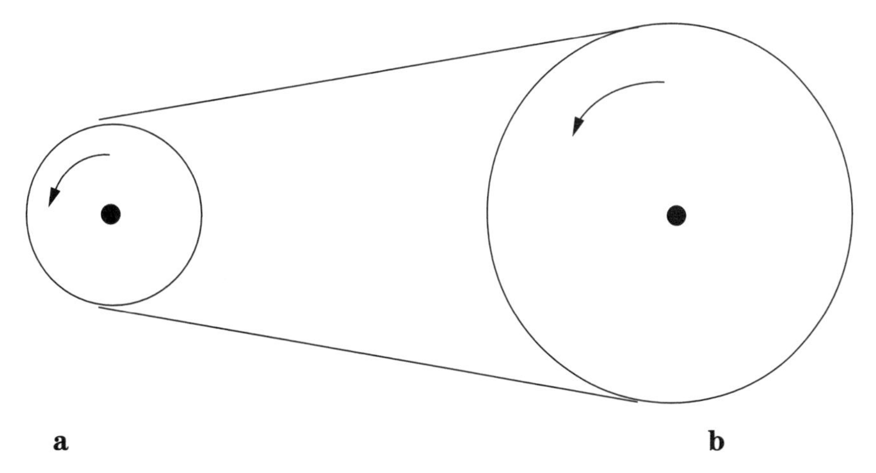

7-4 An open belt drive.

Design rules

The following design rules should be followed when designing drives using round belts of Neoprene or Buna-Nitrile:

- For speeds up to 2500 rpm, pulley diameters should be at least four times the cross-sectional diameter of the belt.
- For speeds above 2500 rpm, pulley diameters should be at least eight times the cross-sectional diameter of the belt.
- The maximum belt speed should be kept below 800 in/sec.
- The center-to-center pulley spacing should not exceed five times the diameter of the larger pulley.
- The minimum cross-sectional diameter for the belt can be determined using:

$$\text{belt diameter} \geq 2.132\sqrt{\frac{\tau}{d_s e}}$$

or

$$\text{belt diameter} \geq 0.412\sqrt{\frac{P}{d_s f_s e}}$$

where

τ = shaft torque, in•oz
d_s = diameter of smaller pulley, in.
f_s = speed of smaller pulley, rpm
P = power transmitted by belt, ft•lb/sec
e = elongation of belt, percent

Sources

The CONSTRUX™ building system toys by Fischer-Price include assorted sizes of ³⁄₃₂-inch-cross-section round belts and the mating grooved pulleys. Large O-rings sold for plumbing use can be used as small round belts. (See chapter 9 for more on

O-rings.) For nondemanding applications, a belt with a square cross section can be used in place of a round belt.

Heavier belt drives

Round belts are fine for low power applications, but for big jobs heavy-duty belt drives are needed. These drives make use of either *flat belts* or *V-belts*.

Flat belts

Flat belts are used in applications requiring (relatively) small pulley diameters, high belt speeds, and low noise. In the past, flat belts were often made of leather, but modern flat belts for heavy use are constructed with multiple plies as shown in Fig. 7-5. A *friction* ply made of synthetic rubber, polyurethane, or sometimes chrome leather forms the belt surface that is in contact with the pulley. The friction ply is laminated to the *tension* ply which is made of drawn polyamide strips or polyester cord. For lower-power applications, flat belts are often made from a nylon or dacron weave that is impregnated with rubber. For operation over very small pulleys (diameters down to 0.05 in.) flat belts are made from mylar sheeting.

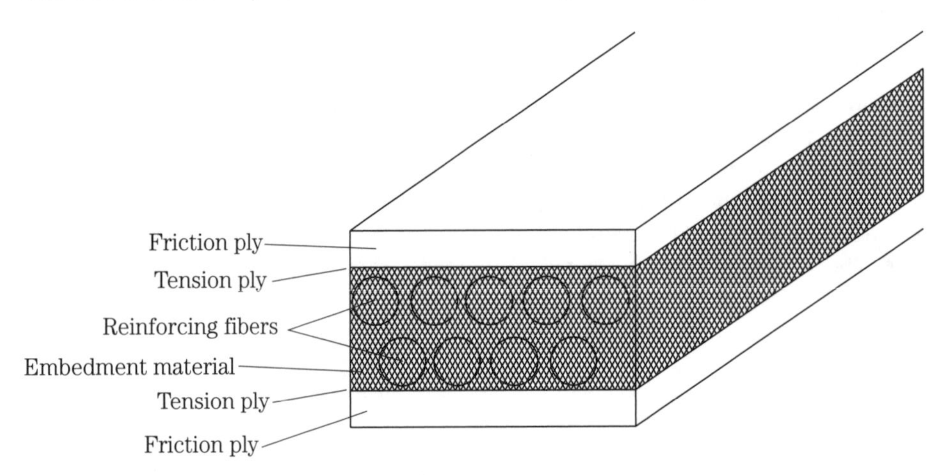

7-5 Cross section of a modern flat belt showing plies.

V-belts

As their name implies, *V-belts* have a V-shaped cross section as shown in Fig. 7-6. Because of their greater thickness, V-belts are not as flexible as flat belts, and therefore they will usually require a larger minimum pulley diameter.

Sources

The best source for heavier flat belts and V-belts is the local automotive parts store. Very small flat belts can often be found in scrapped cassette players and VCRs. For the experimenter, low-elasticity rubber bands are one ready source of light duty flat belts in a variety of sizes. Usually the best rubber bands for use as drive belts—

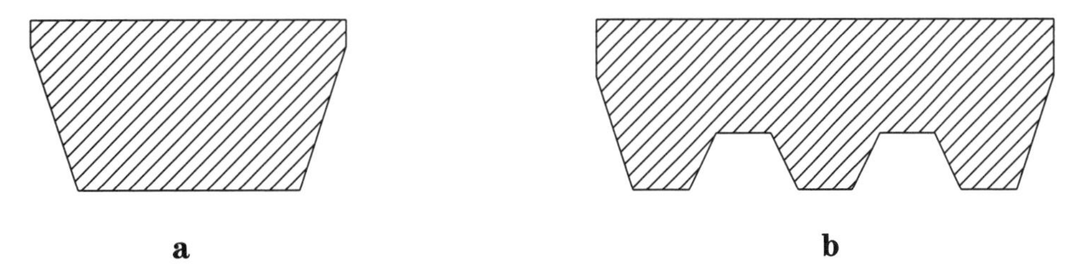

7-6 Cross section of V-belts: (a) single, (b) multiple.

the synthetic, hard to stretch bands that often come in brightly colored assortments—are the worst rubber bands for general office use. The plain brown natural rubber bands favored by offices stretch too easily to make good drive belts.

Design analysis

The configuration shown in Fig. 7-4 is called an *open belt* drive. The *angle of wrap* is the angle over which the belt is in contact with the circumference of the pulley. If both pulleys are of equal size, the belt contacts each pulley over 180 degrees or π radians of circumference. When the pulleys are of unequal size, the belt will contact the smaller pulley for less than 180° and the larger pulley for more than 180°. The angles of wrap for an open belt drive can be determined from the radii of the pulleys and their center-to-center spacing:

$$\alpha_L = 180 + 2 \sin^{-1} \frac{R - r}{C}$$

$$\alpha_S = 180 - 2 \sin^{-1} \frac{R - r}{C}$$

where

α_L = angle of wrap for large pulley, degrees
α_S = angle of wrap for small pulley, degrees
R = radius of large pulley
r = radius of small pulley
C = center-to-center spacing of pulleys

The configuration shown in Fig. 7-7 is called a *crossed belt* drive. The angles of wrap for a crossed belt drive can be determined using

$$\alpha_S = \alpha_L = 180 + 2 \sin^{-1} \frac{R + r}{C}$$

Pulley sizing

One of the fundamental decisions in the design process involves the selection of pulley sizes. The relative sizes of the pulleys are determined by the desired speed ratio:

$$\frac{\omega_S}{\omega_L} = \frac{D_L}{D_S}$$

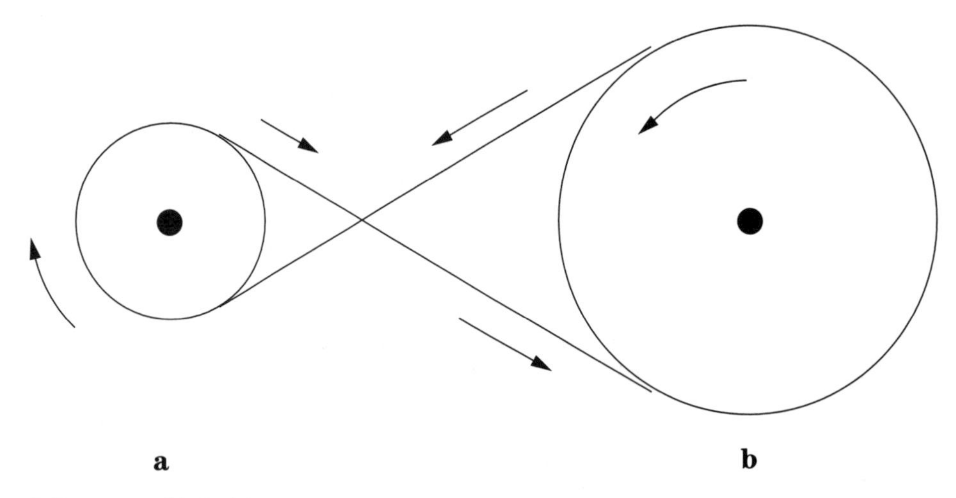

7-7 A crossed belt drive.

where

ω_S = rotational speed of the small pulley
ω_L = rotational speed of the large pulley
D_S = diameter of the small pulley
D_L = diameter of the large pulley

The absolute sizes of the pulleys are governed by the thickness and flexibility of the belt:

- The small pulley **must** be larger than the minimum bending radius of the belt. Furthermore, the pulley **should** be somewhat larger than this minimum to avoid excessive flexing (and possible early failure) of the belt. A good rule of thumb is to use a pulley diameter at least 30 times larger than the thickness of the belt.
- The pulley should not be much larger than necessary, because the larger the pulleys are, the more power will be wasted just to move the pulley masses and the less power will be available for delivery to the load.

For the belt drive shown in Fig. 7-4 let's assume that pulley **A** is driven in a counterclockwise (CCW) direction by a motor or other prime mover. A load of some sort is driven by the shaft of pulley **B**. The CCW rotation of pulley **A** "pulls" on the top portion of the belt and "pushes" on the bottom portion of the belt. This motion, coupled with pulley **B**'s reluctance to rotate due to the attached load, causes the top side of the belt to be at a higher tension than the bottom side of the belt. The basic rule of belt drive design relates the transmitted power, belt speed, and belt tensions:

$$Power = (T_T - T_L)v \tag{7-1}$$

where

T_T = belt tension on the right side
T_L = belt tension on the loose side
v = belt speed

The role of friction

The amount by which T_T and T_L can differ depends upon the angles of wrap and the coefficients of friction between the belt and pulleys. For a flat belt it has been determined that:

$$\frac{T_T - wv^2/g}{T_L - wv^2/g} = (e^{f\alpha})_{\min} \tag{7-2}$$

where

w = density of the belt material
v = speed of the belt
g = acceleration due to gravity = 32.2 ft/sec^2
f = coefficient of friction between belt and pulley
α = angle of wrap

The "min" on the right-hand side of Eq. 7-2 indicates that the values of f and α are for whichever pulley yields a smaller value of $e^{f\alpha}$. When using Eq. 7-2, be careful to keep all the units consistent.

Example 7-2

Consider the case of an open belt drive consisting of two cast iron pulleys connected by a leather belt that is 2 inches wide and 0.125 inch thick. The diameters of the pulleys are 8 and 24 inches. The center-to-center spacing between pulleys is 30 inches. The belt density is 0.035 lb/in^3, and the coefficient of friction between leather and cast iron is 0.35. The maximum allowable stress for the belt is 300 psi. Assume that the belt is to be operated at maximum power capacity with the smaller pulley driven at 600 rpm. Determine the tensions T_L and T_T along with the maximum power P.

Solution:

- The cross-sectional area of the belt is (2)(0.125) = 0.25 in^2. Therefore, T_T is limited to a value that will not subject the belt to stress in excess of 300 lb/in^2.

$$T_T = 300\frac{\text{lb}}{\text{in}^2} \times 0.25\ \text{in}^2 = 75\ \text{lb}$$

- The angles of wrap are obtained using Eq. 7-2:

$$\alpha_L = 180 + 2\sin^{-1}\frac{R-r}{C}$$
$$= 210.9° = 3.68\ \text{rad}$$
$$\alpha_S = 180 - 2\sin^{-1}\frac{R-r}{C}$$
$$= 149.1° = 2.6\ \text{rad}$$

- Because both pulleys have the same coefficient of friction, the smaller value of $e^{f\alpha}$ will be obtained for the smaller pulley.

$$(e^{f\alpha})_{\min} = e^{(0.35)(2.6)} = 2.484$$

- The speed of the belt is obtained as:

$$v = 600\ \frac{\text{rev}}{\text{min}} \times 2\pi(8)\frac{\text{in}}{\text{rev}} \times \frac{1}{12}\frac{\text{ft}}{\text{in}} \times \frac{1}{60}\frac{\text{min}}{\text{sec}} = 41.9\ \frac{\text{ft}}{\text{sec}}$$

- In order to use Eq. 7-2, we need to compute the quantity wv^2/g and have the result in pounds (so it can be subtracted from T_T, which is already in pounds).

$$\frac{wv^2}{g} = 0.035\frac{\text{lb}}{\text{in}^3} \times 0.25\ \text{in}^2 \times (41.9)^2\frac{\text{ft}^2}{\text{sec}^2} \times \frac{1}{32.2}\frac{\text{sec}^2}{\text{ft}} \times 12\frac{\text{in}}{\text{ft}}$$

$$= 5.725\ \text{lb}$$

- Finally, using Eq. 7-2, we obtain:

$$\frac{T_T - wv^2/g}{T_L - wv^2/g} = \left(e^{f\alpha}\right)_{\min}$$

$$\frac{75 - 5.725}{T_L - 5.725} = 2.484$$

$$T_L = 5.725 + \frac{75 - 5.725}{2.484}$$

$$= 33.6\ \text{lb}$$

- The power is obtained using Eq. 7-1:

$$P = (T_T - T_L)v$$

$$= (75 - 33.6)(41.9) = 1734.7\ \text{lb/sec}$$

Note: 1 horsepower = 550 ft•lb/sec, so this power is equal to 1734.7/550 = 3.15 hp.

Synchronous belts

Synchronous belts or *timing belts* are thick flat belts with teeth cut into their inside surface. These teeth mesh with teeth cut into the circumference of special pulleys called *timing pulleys*. Heavy duty timing belts and pulleys are used on many overhead cam 4-cylinder automobile engines. Such belts are readily available at automotive supply stores, but the timing pulleys must be obtained from a junkyard or ordered (at considerable expense!) through a car dealer's parts department. Smaller timing belts and pulleys can be found in many computer printers and some VCRs. One source for small but tough timing belts is a hobby shop that sells parts for radio-controlled model helicopters.

Gears

The design of gears is a somewhat complicated affair, and there are books intended for mechanical engineers and mechanical designers that cover the subject of gear design in great detail. This chapter will not pretend to compete with these design treatises. The typical amateur experimenter is not likely to be creating custom gear designs. This chapter is intended to equip the reader to make best use of whatever suitable gears can be found for a particular application. Professional designers determine what they need for a particular application and then order or specify the exact gear desired. The home experimenter, on the other hand, is more likely to formulate an approximate design, scrounge some gears that will be "good enough," and then finalize the design of the mechanism around the particular gears that were obtained. Some of the things that experimenters might need to know include:

- The different types of gears and which types are best for which applications
- Different possible types of tooth profile and how to ensure proper meshing of gears
- How to estimate tooth strength
- Combining gears into gear trains
- How to put gears onto shafts
- How to properly position gear shafts within a frame

Spur gears

When talk turns to gears, most people form a mental image of spur gears. An assortment of spur gears is shown in Fig. 7-8. A closeup view of several of these gears is shown in Fig. 7-9. Notice that there are several different tooth shapes or *profiles.* One gear has relatively pointed *troughs* between its teeth, but the tips of the teeth are slightly flattened. Another gear has troughs with noticeably squared-off bottoms, and the third has pointed teeth with almost no noticeable flattening.

The teeth on gears are specifically designed to ensure proper contact and maximum transfer of power between mating gears. Theoretically, there are many different gear-tooth profiles that can be used. In general, with certain very loose re-

7-8 Assorted spur gears.

7-9 Closeup of spur gears showing different tooth profiles.

strictions, it is possible to design one gear with an arbitrary set of tooth profiles and then design a mating gear that will mesh with the first gear. The second gear's tooth profiles are not arbitrary, and in general, they will be different from the first gear's tooth profiles. If we start with an arbitrary gear design, say **gear A**, and design two other gears, **gear B** and **gear C**, each of which meshes with **gear A**, then in general, **gear B** will not mesh with **gear C**. There are some specific tooth profiles designed to make **gears B** and **C** mesh with **gear A** and also allow **gears B** and **C** to mesh with each other. Such profiles are called *interchangeable* gear tooth forms.

A naive design for spur gears

Let's explore some of the issues involved in gear specification and design by making a pair of simple spur gears out of cardboard. The larger gear will have 16 teeth and be approximately 3 inches in diameter, and the smaller gear will have 12 teeth. The diameter of the smaller gear will be determined as part of the design process.

We begin by drawing a circle 3 inches in diameter and dividing the circumference into 16 equal parts as shown in Fig. 7-10. This circle will be the *pitch circle* of the larger gear. The desired gear will have one tooth located at each of the 16 tic marks shown. To have exactly 16 teeth evenly spaced around the circle, one tooth and one space must fit exactly into each 22.5° wedge as shown in Fig. 7-11. Notice that the sides of the teeth extend both inside (or below) and outside (or above) the pitch circle, but corresponding edges of consecutive teeth cross the pitch circle exactly 22.5° apart.

The space between teeth must be wide enough to accommodate a tooth on the mating gear. If we assume that both gears will have the same tooth width, this means that each tooth on the larger gear can take up no more than half of the available 22.5°

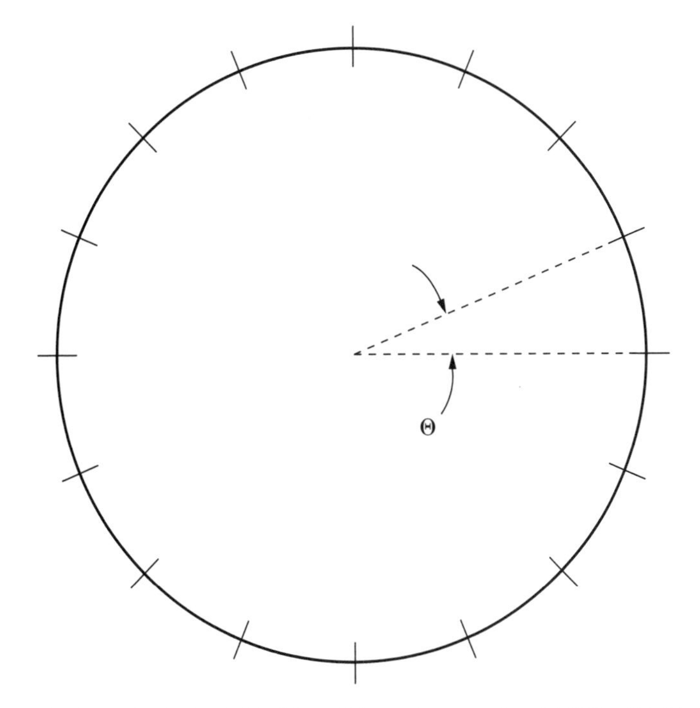

7-10 Pitch circle divided into 16 equal arcs of $\theta = 360/16 = 22.5°$.

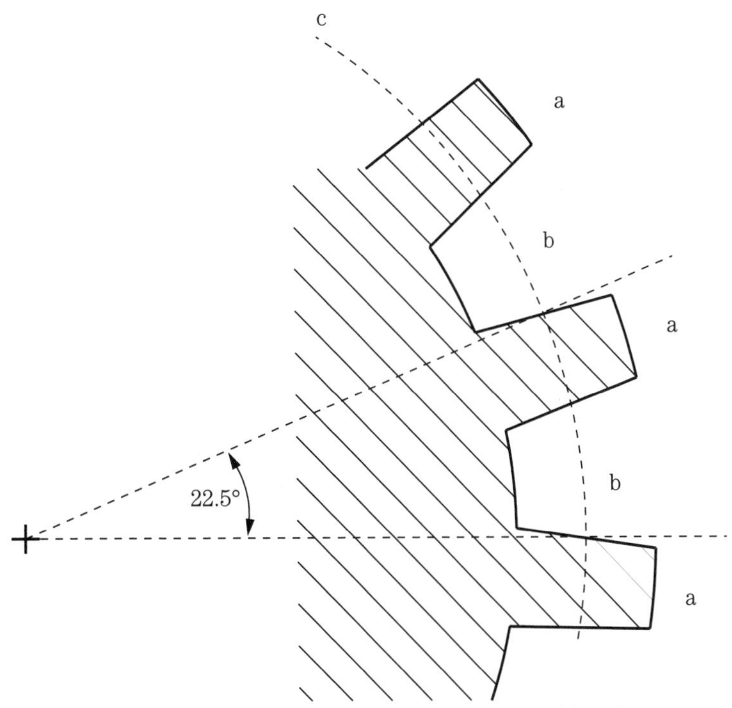

7-11 Enlarged section of gear showing how one tooth (a) and one space (b) fit exactly in one 22.5° arc of the pitch circle (c).

wedge. Let's arbitrarily make each tooth 9.5° wide and each space 13° wide. We then add tic marks 9.5° clockwise from each of the tics shown in Fig. 7-10 to obtain the circle shown in Fig. 7-12. Although we have fixed the widths of teeth and spaces as measured at the pitch circle, we still must decide about the slope of the teeth sides. One possibility is to let each tooth side lie on the same line at a radius of the pitch circle. This approach results in "square" teeth as shown in Fig. 7-13.

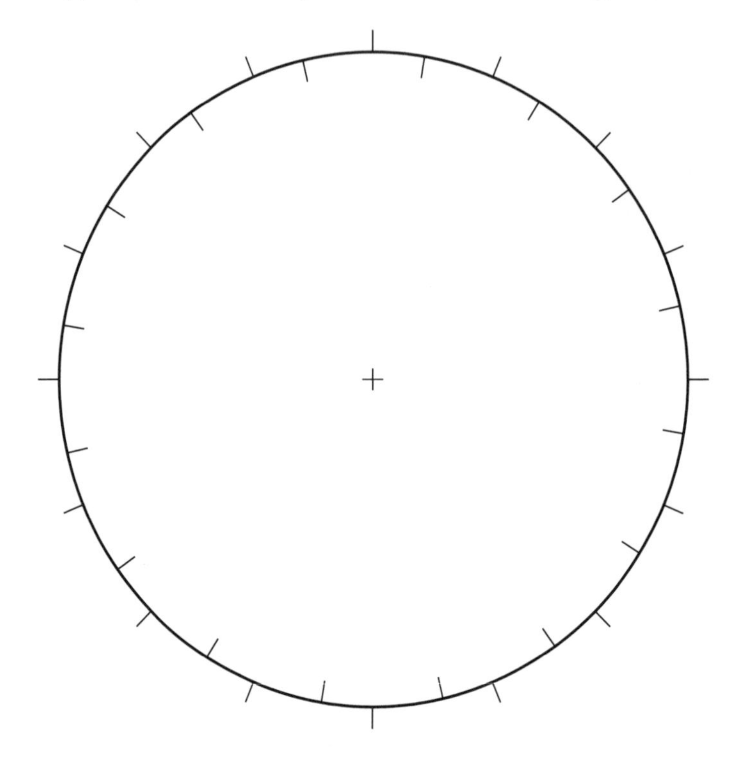

7-12 Pitch circle divided into alternating arcs of 9.5° and 13°.

Gears with square teeth will run very roughly if they run at all. On the other hand, if the tooth sides slope too gently, the depth of the spaces becomes very shallow as shown in Fig. 7-14, and the mesh becomes prone to slippage. We can obtain something that looks reasonable if we incline the tooth sides ±20° from radii of the pitch circle as shown in Fig. 7-15. At this point, we could decide to stop working on the large gear and move on to design the smaller gear to mesh with the profile shown in Fig. 7-16. However, if we were to actually construct a gear using the profile shown, the pointed tips of the teeth would be very weak and prone to bend or break.

Instead, we can draw two circles which are concentric with the pitch circle and which serve to truncate the height of the teeth and depth of the spaces as shown in Fig. 7-16. The circle that defines the tops of the teeth is called the *outer circle*, and the circle that defines the bottoms of the spaces is called the *root circle*. The radial distance between the pitch circle and the outer circle is called the *addendum.* The radial distance between the pitch circle and the root circle is called the *dedendum.* (The terms addendum and dedendum are sometimes used to designate the parts of

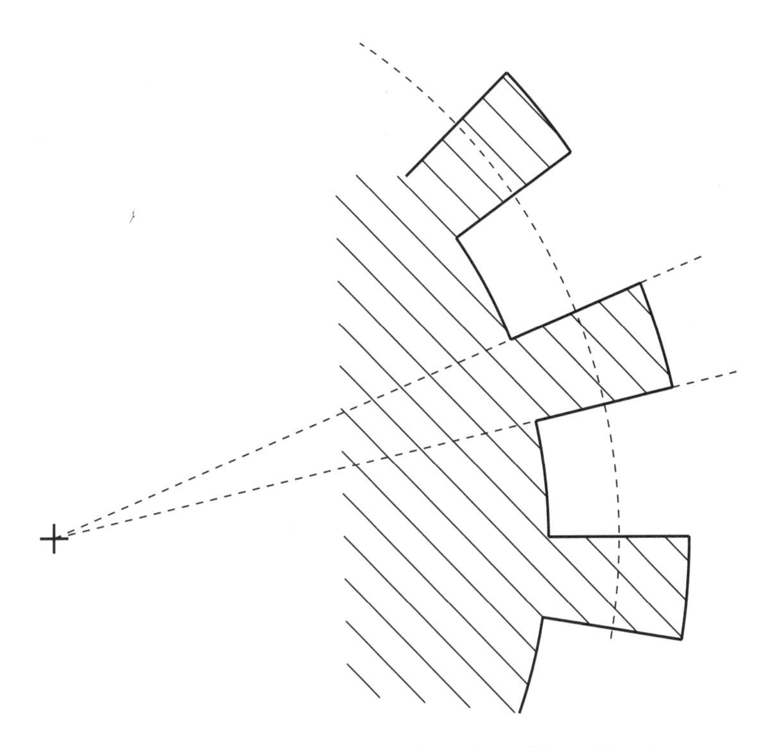

7-13 Tooth sides coincident with pitch-circle radii resulting in square teeth.

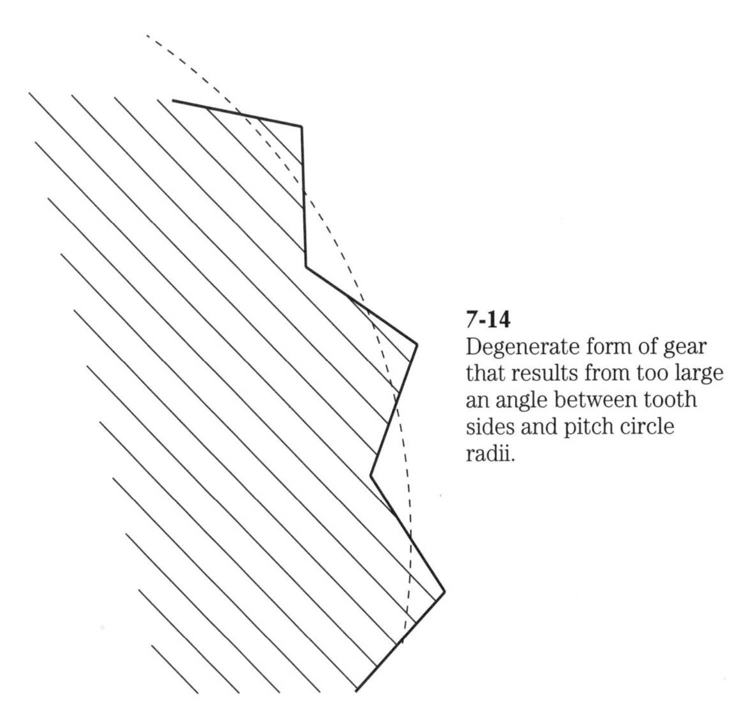

7-14
Degenerate form of gear that results from too large an angle between tooth sides and pitch circle radii.

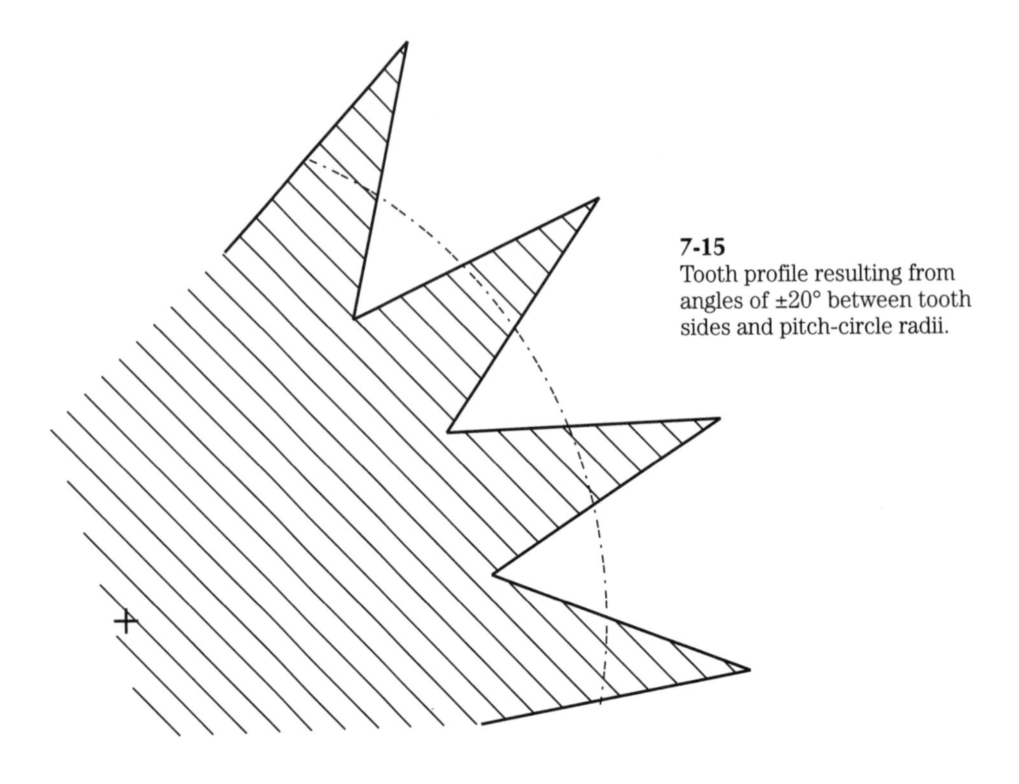

7-15
Tooth profile resulting from angles of ±20° between tooth sides and pitch-circle radii.

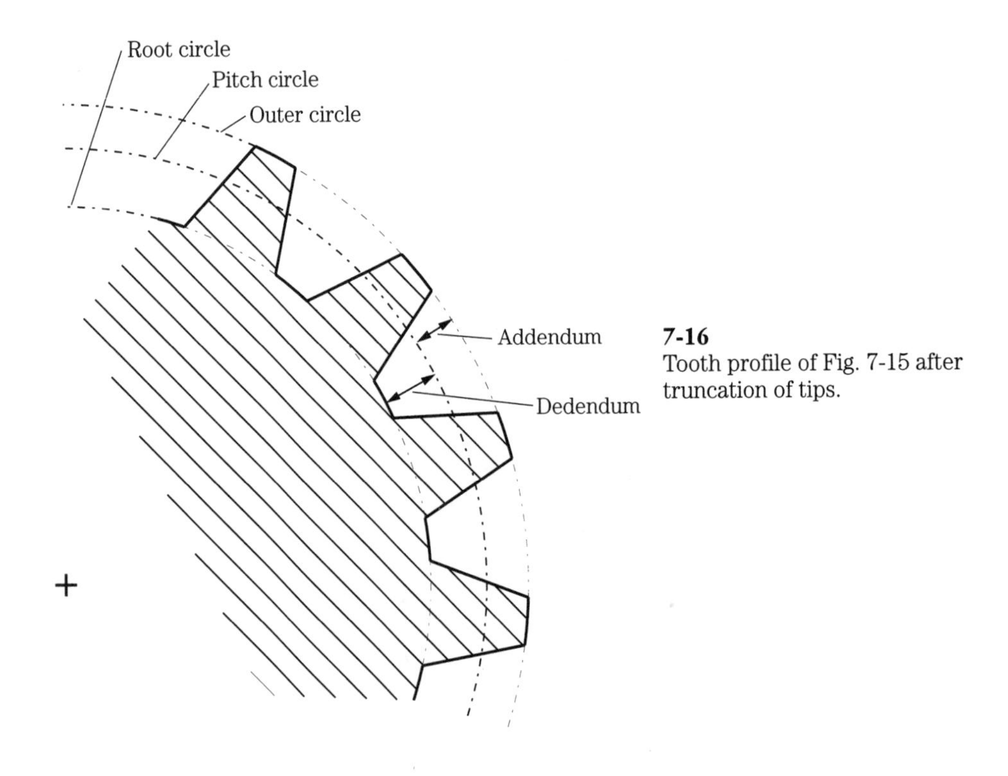

7-16
Tooth profile of Fig. 7-15 after truncation of tips.

the teeth lying above and below the pitch circle.) Let's choose an outer circle diameter of 3.3 inches and a root circle diameter of 2.6 inches. The resulting 16-tooth gear is shown full size in Fig. 7-17.

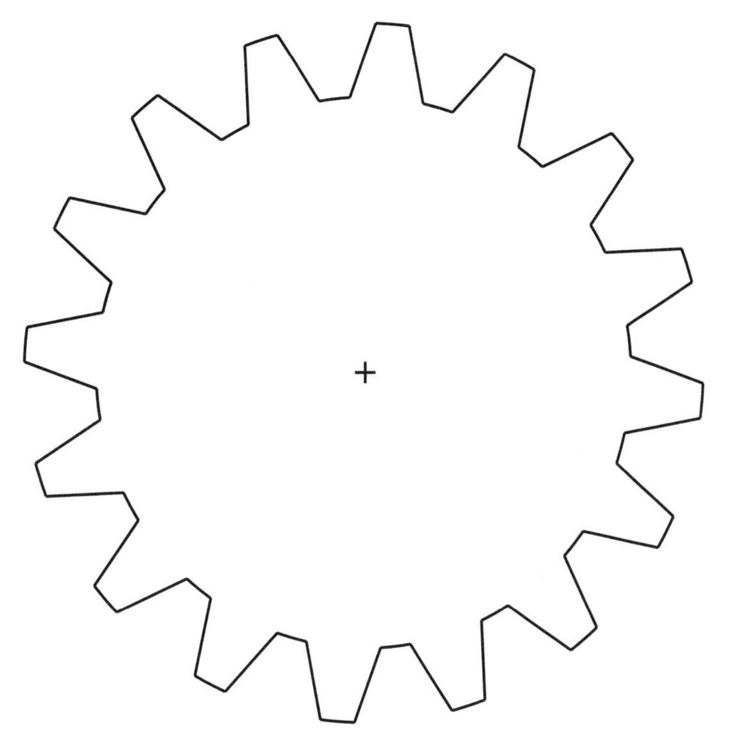

7-17 Spur gear with 16 straight-sided teeth.

It is a fundamental rule of gear design that within a family of spur gears designed to mesh with each other, the ratios of pitch-circle diameters must equal the corresponding ratios between numbers of teeth. (This rule is an unavoidable consequence of the obvious requirement that every gear have an integer number of teeth and that these teeth must be evenly spaced around the pitch circle's circumference.) For our current example, we want a tooth-count ratio of 16:12 or 4:3. Therefore, we find the diameter of the smaller gear's pitch circle as:

$$\frac{d}{3} = \frac{3}{4}$$

$$d = \frac{9}{4} = 2.25 \text{ in.}$$

We proceed by drawing a pitch circle having a diameter of 2.25 inches and dividing the circumference into 12 equal arcs of 30° each. We decided earlier that the two gears should have teeth of equal width as measured along the pitch circles. This doesn't mean that we make the teeth fit within a 9.5° wedge as we did with the first gear. What it does mean is that we must first determine the width (as measured along the 3-inch pitch circle) of a tooth on the first gear, and then determine the angle subtended by an identical distance on a 2.25-inch pitch circle.

A tooth's width as measured along the pitch circle is given by the formula:

$$p = \frac{\pi \theta d}{360} \tag{7-3}$$

where

θ = width of tooth in degrees
d = diameter of the pitch circle

Therefore, an arc of 9.5° on a 3-inch circle has length of:

$$\frac{3.14159(9.5)(3)}{360} = 0.2487$$

We can then use an inverted form of Eq. 7-3 to determine the width in degrees of a tooth having a width of 0.2487 in. as measured on a 2.25-inch pitch circle.

$$\begin{aligned} \theta &= \frac{360p}{\pi d} \\ &= \frac{(360)(0.2487)}{(3.14159)(2.25)} \\ &= 12.66° \end{aligned}$$

Alternatively, for the case of equal widths along the pitch circle, we can use Eq. 7-3 twice to obtain:

$$p_1 = \frac{\pi \theta_1 d_1}{360} = p_2 = \frac{\pi \theta_2 d_2}{360}$$

which in terms of angular widths and diameters reduces to:

$$\frac{\theta_2}{\theta_1} = \frac{d_1}{d_2} \tag{7-4}$$

A quick check confirms that for $d_1 = 3.0$, $d_2 = 2.25$, and $\theta_1 = 9.5$; Eq. 7-4 yields $\theta_2 = 12.666$ as expected. Using the same addendum (0.15) and dedendum (0.2) as used for the 16-tooth gear, we obtain the 12-tooth gear shown in Fig. 7-18.

Mesh conditions

Let's take a closer look at the mesh between the two gears we have designed and see why this sort of tooth profile is not in common use. A closeup view of the mesh area is shown in Fig. 7-19. Assume that the 12-tooth gear is the driver rotating counterclockwise. At the instant depicted in Fig. 7-19, the corner of one tooth on the smaller gear is about to dig into the face of one tooth on the larger gear. In addition to the potential gouging or binding, this picture illustrates another problem with this tooth profile, namely uneven rotation. As shown, the tip of the smaller gear's tooth is contacting the larger gear at a point approximately 1.35 inches from the center.

As the gears rotate, the point of contact will slide out along the tooth of the larger gear until the contact is approximately 1.65 inches from the center of the larger gear.

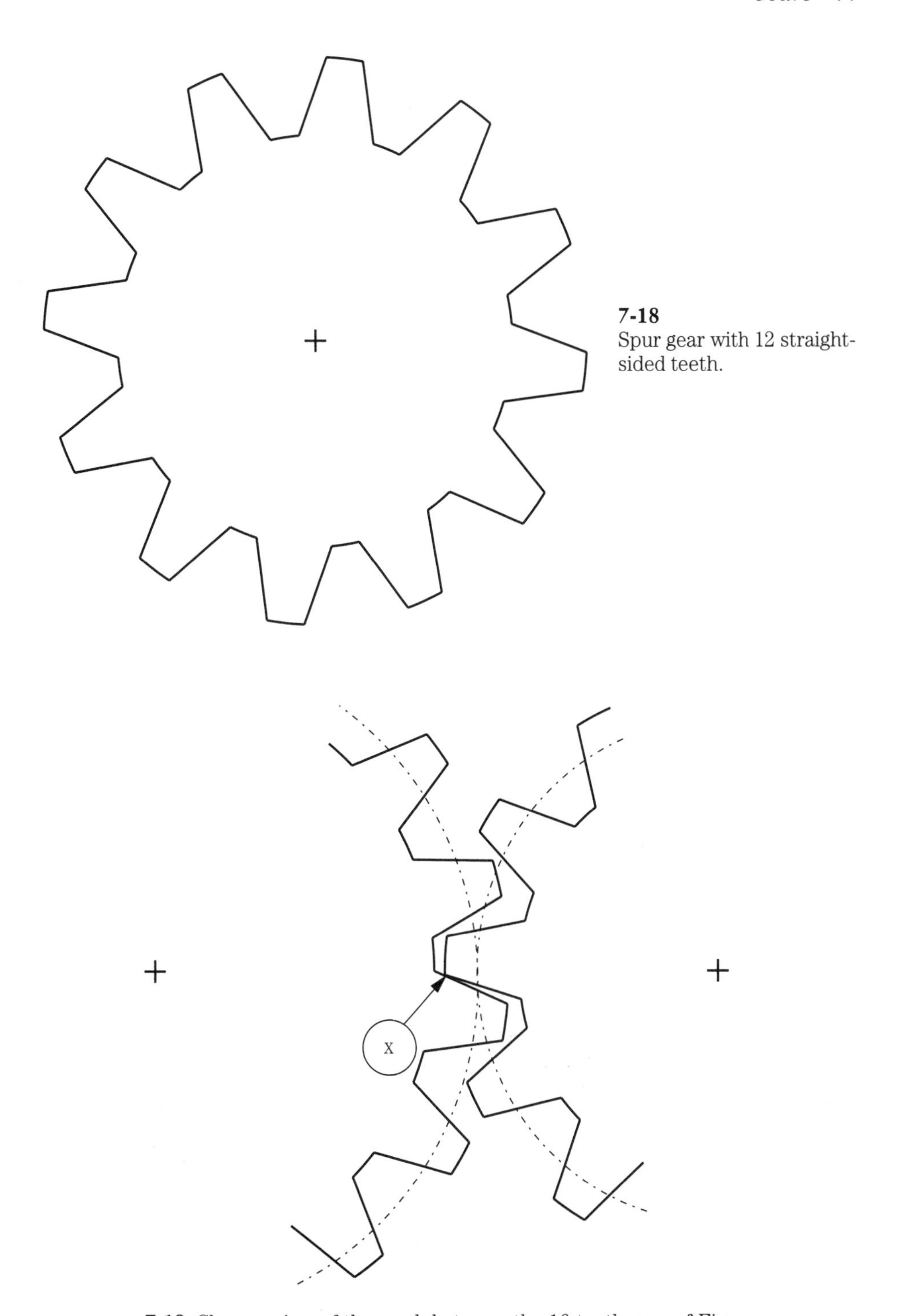

7-18
Spur gear with 12 straight-sided teeth.

7-19 Closeup view of the mesh between the 16-tooth gear of Fig. 7-17 and the 12-tooth gear of Fig. 7-18. The X indicates a site of possible binding, gouging, or chipping.

The smaller gear has an outer diameter of 2.55 inches, and the tips of the teeth therefore move along a circular path having a total length of 2.55π inches. Let's assume that the small gear is rotating at one rpm. Then the tip of each tooth is moving along the outer circle at a rate of 2.55π inches per minute. At the instant depicted in Fig. 7-19, the tip of the smaller gear's tooth is contacting the larger gear on the circumference of an imaginary circle having a radius of 1.35 inches. Along this circle, the tooth of the larger gear is being driven at an instantaneous rate of 2.55π inches per minute. Because the circumference of the imaginary circle is 2.7π, the larger gear's instantaneous rotational speed must be

$$\frac{2.55\pi}{2.7\pi} = 0.944 \text{ rpm}$$

However, when the point of contact has slid out to a point 1.65 inches from the center of the larger gear, the imaginary circle (on the larger gear) passing through the point of contact has a circumference of 3.3π inches. Therefore, the larger gear's instantaneous rotational speed must be

$$\frac{2.55\pi}{3.3\pi} = 0.7727 \text{ rpm}$$

The speed of the larger gear varies as the points of contact move in and out along the tooth of the larger gear.

The potential for binding and the speed variation exposed above both indicate that straight-sided teeth do not result in a particularly good gear design. What we need are tooth profiles curved in such a way that the tooth faces on one gear contact the tooth faces on the other gear at points that move smoothly along the tooth face without binding and without causing a speed variation. The *involute* curve is one curve that satisfies all the constraints needed to make this smooth constant-speed operation possible. We will explore the properties of involute curves after we look at what it takes to provide a constant speed ratio between the driving and driven gears.

Conjugate action

Consider the case of two meshed gears. The input gear or *driving* gear is turned by some external mover, let's say for this example a motor connected to the gear's shaft. The output gear or *driven* gear is driven by the contact made between the teeth of the two gears. Let's assume that the input gear is rotating at a constant speed. In order to maintain a constant rotational speed of the output gear, the tooth profiles must obey what is called the *fundamental law of gearing*: for the velocity ratio of the two gears to remain constant, the shape of their contacting profiles must be such that a common line normal to both profiles at the point of contact passes through a fixed point on the *line of centers.* As shown in Fig. 7-20, this fixed point, called the *pitch point,* lies at the common intersection of the line of centers and the pitch circles of both gears. The angle between the normal common to both profiles and the tangent common to both pitch circles is called the *pressure angle.* (Common pressure angles include 14½°, 20°, 22½°, and 25°.) When the fundamental law of gearing is satisfied for all rotational positions of two gears, the teeth of the two gears are said to have *conjugate profiles*, or to exhibit *conjugate action.* Two tooth profiles that yield interchangeable gear forms and that satisfy the fundamental law of gearing are the *involute* and *cycloidal* profiles.

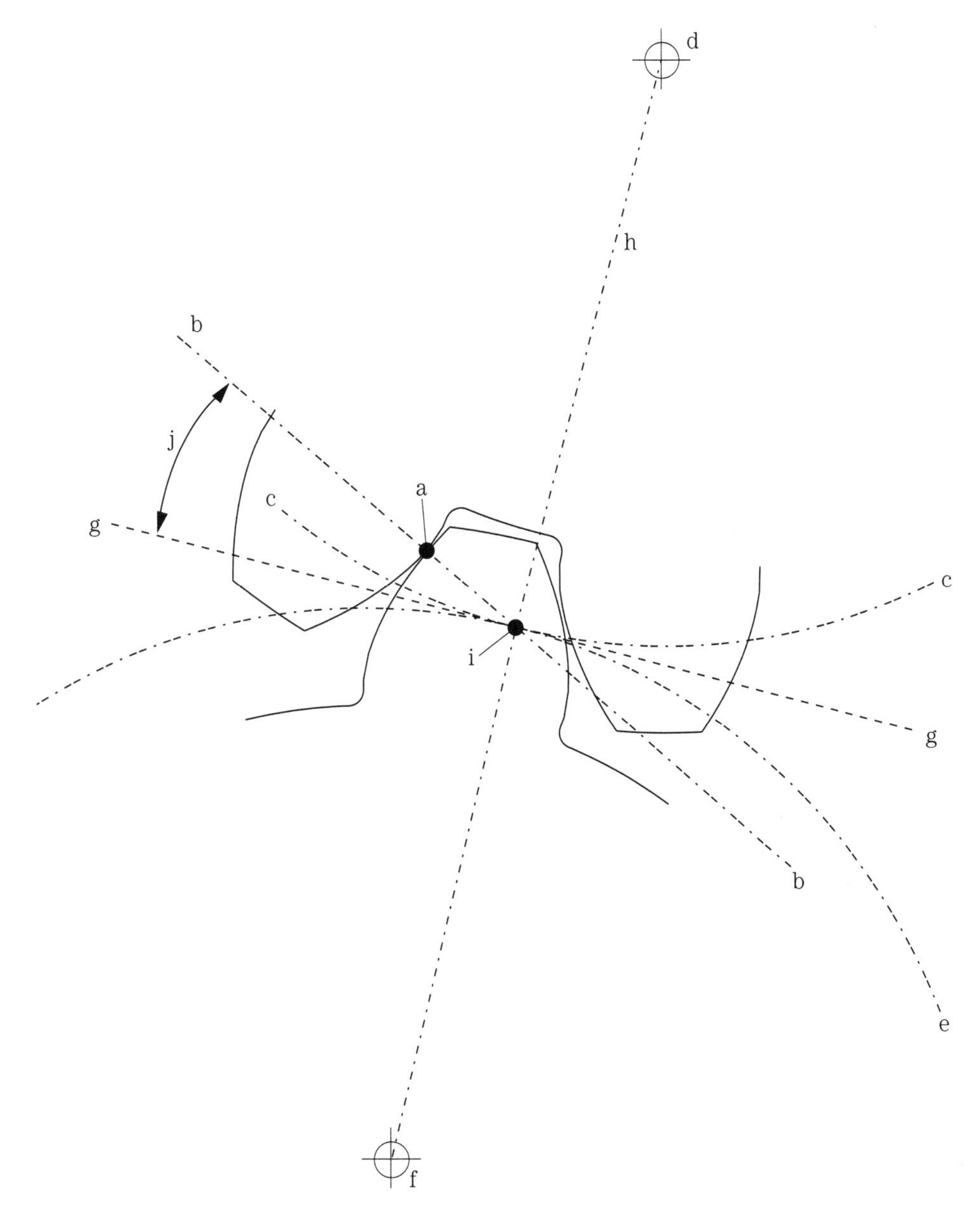

7-20 Geometry of the fundamental law of gearing: (a) point of contact, (b) common normal, (c) pitch circle of top gear, (d) center of top gear, (e) pitch circle of bottom gear, (f) center of bottom gear, (g) common tangent to both pitch circles, (h) line of centers, (i) pitch point, (j) pressure angle.

Involute curves

Let's start with a simple experiment. Take a circular disc and wrap a string around the edge as shown in Fig. 7-21a. The last point at which the string contacts the disc is labeled **A** in the figure. If we slowly unwind the string, the point **A** on the string will describe a curve as shown in Fig. 7-21b. The curve is the involute curve that is used to specify gear-tooth profiles.

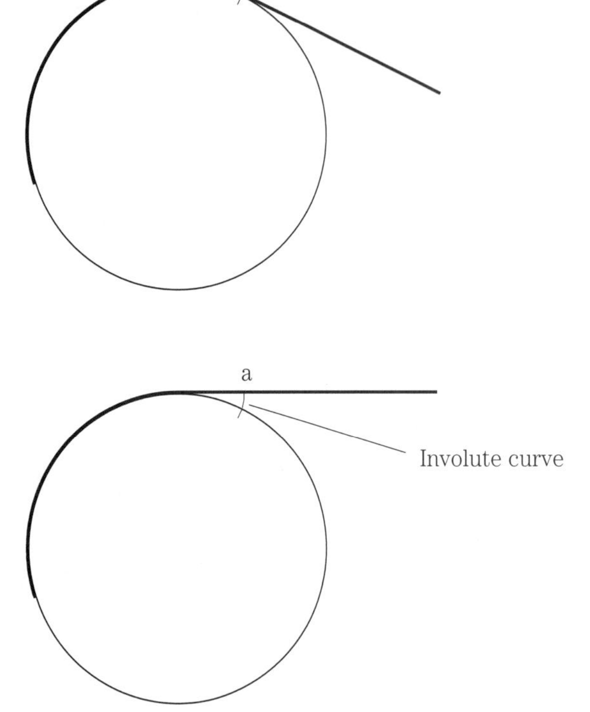

7-21
Graphical construction of an involute.

Mathematical formula for generating an involute

Let A denote the point on the involute being generated at one instant in the unwrapping process described above. A snapshot of the generating geometry at this instant is depicted in Fig. 7-22. The first point on the involute is denoted A_0. The angle between the radius to A_0 and the radius to the point of tangency T is denoted as ϕ. If we can obtain a mathematical expression for the position of point A as a function of ϕ, we can use a computer to calculate and plot the position of A as ϕ is swept from zero through an angle sufficient to generate half the tooth profile. It is a simple matter to use plotting software to manipulate this piece of involute as needed to draw an entire tooth or an entire gear.

Development of the desired function is a straightforward process of using the information that can be extracted from Fig. 7-22.

- The use of polar coordinates seems naturally suited for this particular application. Let the center of the base circle be the origin and let the angle to point A_0 be zero. Then the coordinates of point A are $(r = R_B, \phi = 0)$ where R_B is the radius of the base circle.
- The arc length measured along the base circle from point T to point A_0 is given by

$$s = \frac{\phi \pi R_B}{180}$$

- This length is also the straight-line distance measured along the tangent from point T to point A.
- As shown in the figure, the radius to T and the length s along the tangent form two sides of a right triangle. The length of this triangle's hypotenuse is the polar-coordinate distance from the origin to point A, and the angle between the hypotenuse and the radius to T is the polar coordinate angle to point A.
- Using the Pythagorean theorem, the length of the hypotenuse is obtained as:

$$\begin{aligned} r &= \sqrt{R_B^{\,2} + s^2} \\ &= \sqrt{R_B^{\,2} + \left(\frac{R_B \pi \, \phi}{180}\right)^2} \qquad (7\text{-}5) \\ &= R_B \sqrt{1 + \left(\frac{\pi\phi}{180}\right)^2} \end{aligned}$$

- The angle between the radius to T and the hypotenuse is given by:

$$\rho = \tan^{-1} \frac{s}{R_B} = \tan^{-1} \frac{\phi\pi}{180}$$

The polar angle from A_0 to A is then obtained as the difference between ϕ and ρ.

$$\theta = \phi - \rho = \phi - \tan^{-1} \frac{\phi\pi}{180} \qquad (7\text{-}6)$$

Equations 7-5 and 7-6 constitute a set of parametric equations in ϕ for the polar coordinate radius r and angle θ. These coordinates can be plotted directly using plotter software that handles polar coordinates, or the polar coordinates can be converted into rectangular coordinates using the following formulas:

$$x = r \cos \theta$$
$$y = r \sin \theta$$

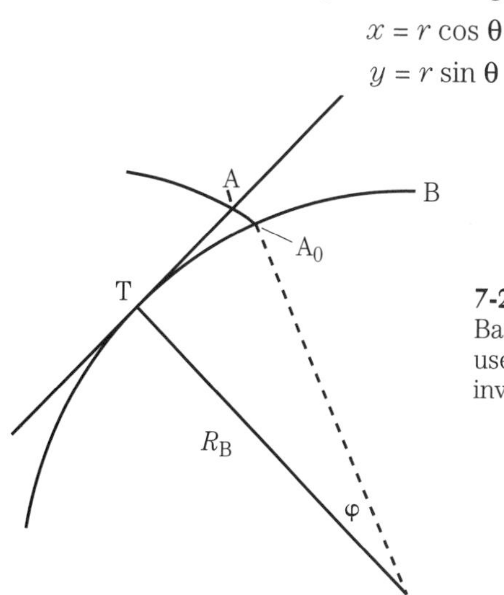

7-22
Base circle with tangent and radii used to generate a point on the involute curve starting at point A_0.

Constructing an involute gear profile

1. Determine the desired pitch diameter, and draw the pitch circle.
2. Draw a radius of the pitch circle. The point at which the radius intersects the circle will become the pitch point.
3. Draw a tangent to the pitch circle at the pitch point.
4. Determine the desired pressure angle. Draw a line that passes through the pitch point and makes the desired pressure angle with the tangent drawn in step 3. This line will be called the pressure line.
5. Draw a line that passes through the center of the pitch circle and that is perpendicular to the pressure line.
6. The point at which the line drawn in step 5 intersects the pressure line defines the radius of the base circle. Draw the circle that passes through this point and that has the same center as the pitch circle. This second circle is the base circle.
7. For a gear with N teeth, mark off N equal arcs on the circumference of the base circle.
8. In conjunction with the base circle drawn in step 6, construct an involute curve at each of the marks made in step 7.

Bevel gears

Although in prior sections we often talked about spur gears as though they were built around circles in a two-dimensional plane, in reality spur gears are built around three-dimensional cylinders. Imagine now that instead of cutting teeth into the surface of a cylinder, we cut teeth into the surface of a cone and cut off the tip of the cone. We would obtain a bevel gear like the one shown in Fig. 7-23.

Bevel gears are used on shafts whose axes intersect as shown in Fig. 7-24. The case depicted, where the axes meet at a 90-degree angle, is probably the most common situation, but bevel gears can also be made for other angles. Some examples are shown in Fig. 7-25.

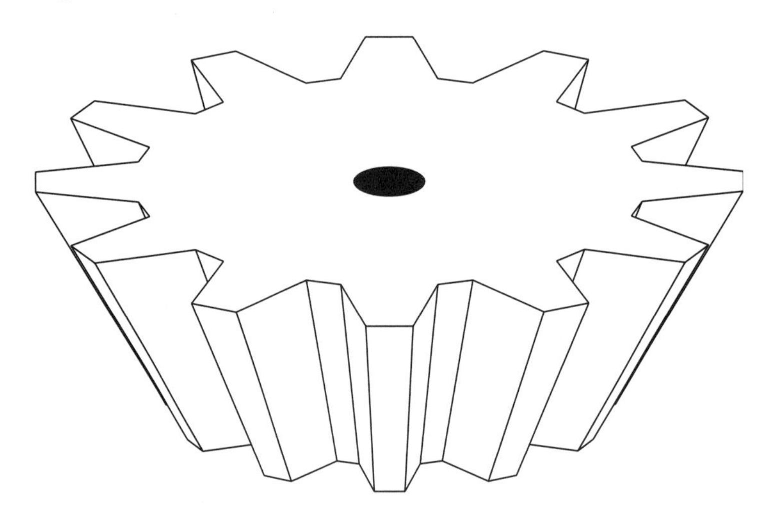

7-23 Bevel gears.

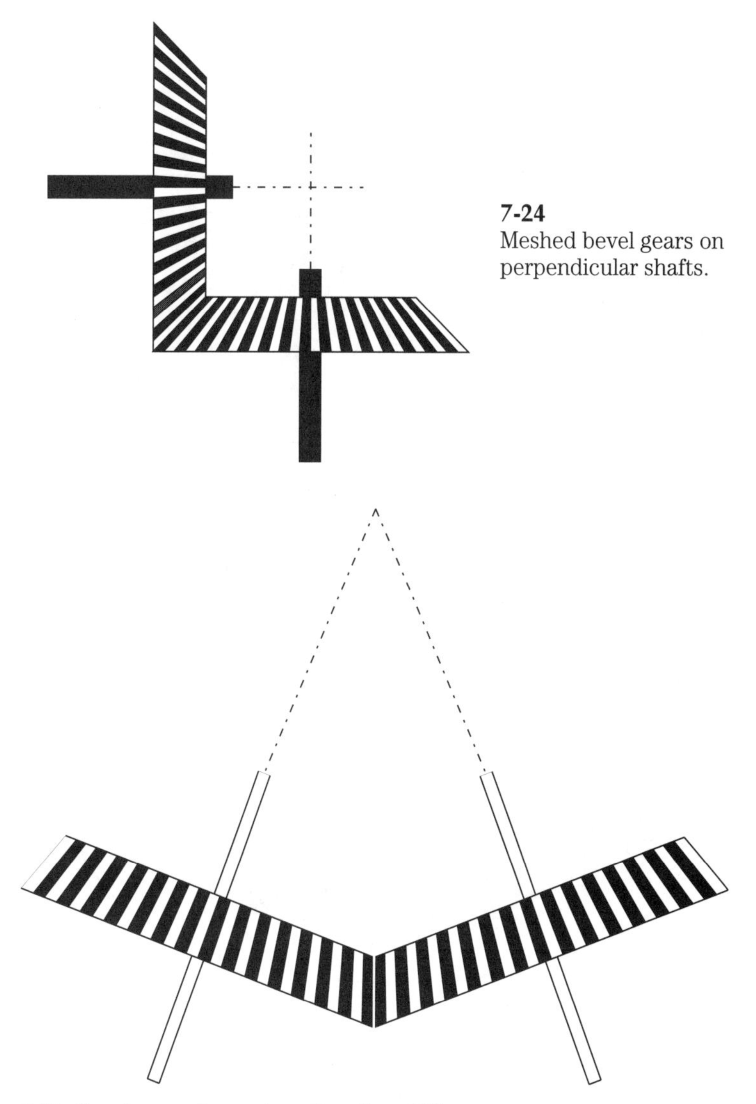

7-24
Meshed bevel gears on perpendicular shafts.

7-25 Bevel gears for angles other than 90°.

Face and crown gears

A face gear is a circular disc with teeth cut into one of the flat circular faces as shown in Fig. 7-26. A similar type of gear called a crown gear is shown in Fig. 7-27. Such a gear can be used in conjunction with a spur pinion as shown in Fig. 7-28 to perform essentially the same function that is performed by a pair of bevel gears each having a pitch angle of 45 degrees.

Figure 7-29 is a photograph of a gear that is part of the LEGO Technics building toy system. This gear is actually a hybrid design; part spur gear, part crown gear. The shaft axes of two of these gears can be positioned at any angle from parallel, through 90°, to about 315° as shown in Fig. 7-30.

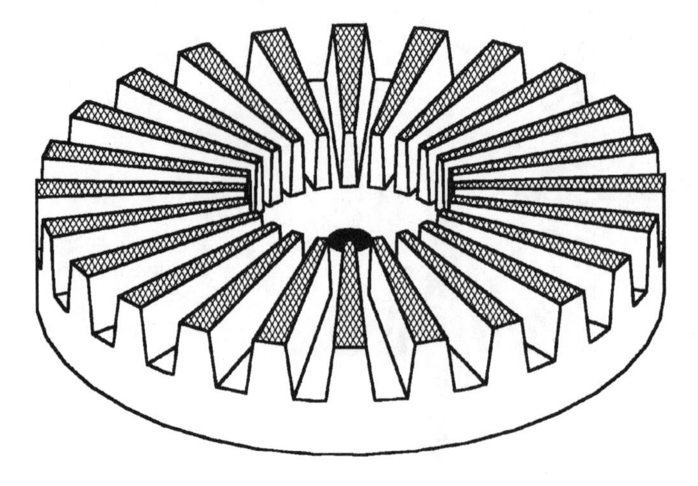

7-26 Face gear.

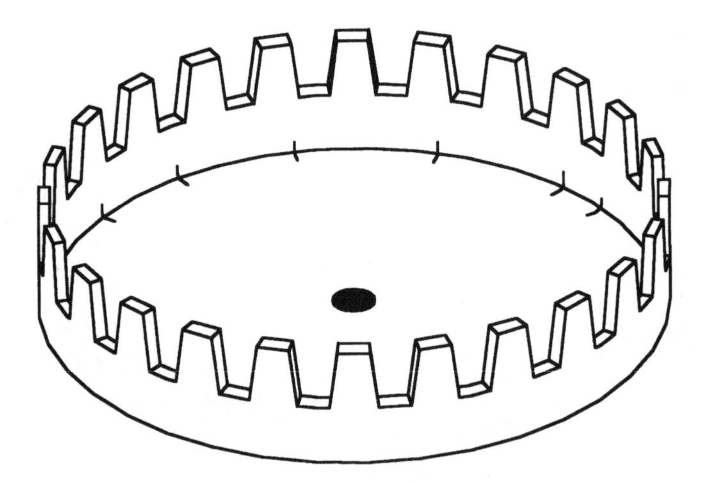

7-27 Crown gear.

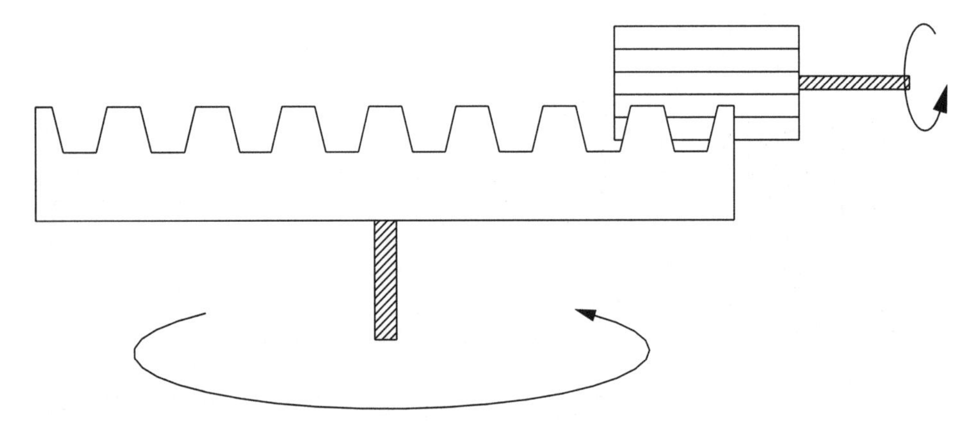

7-28 Pinion gear in mesh with crown gear.

7-29
Hybrid spur-crown gears from LEGO building set.

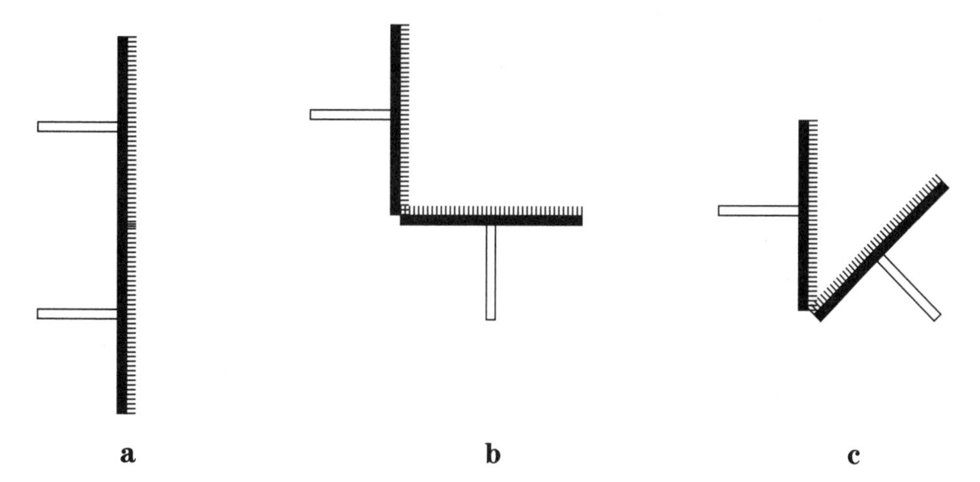

7-30 The hybrid gears from the LEGO building set can have their shafts at a range of angles: (a) parallel, (b) perpendicular, (c) 315 degrees.

Worm gears

A *worm* and *worm wheel* are shown in Fig. 7-31. The worm resembles a section of coarse-threaded bolt. The worm wheel resembles a spur gear and is sometimes called the *worm gear*. The axes of the worm and worm wheel do not intersect, and they are usually orthogonal. A cross-sectional view of a worm gear is shown in Fig. 7-32. Unlike a spur gear, the width of the teeth tapers towards the tips and the tops of the teeth are made concave to match the curve of the worm.

The taper on the teeth is usually laid out such that if the teeth were extended, they would come to a sharp point at the center axis of the worm as shown in Fig. 7-33. The teeth on a worm wheel are slanted much like the teeth on a helical gear. The helix angle of the worm wheel is matched to the lead angle on the worm. For light-duty applications, if the lead angle of the worm is small, an ordinary spur gear of the proper pitch can be used instead of a helical worm wheel. Some amateur ro-

7-31 Worms and worm wheels.

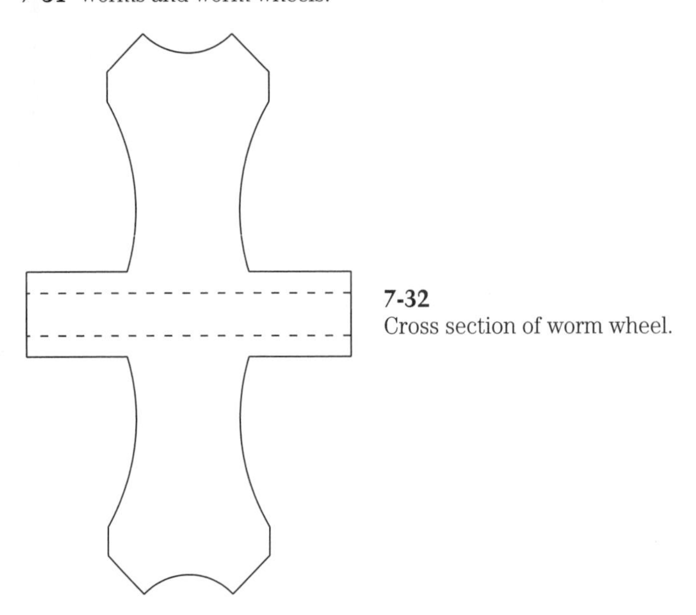

7-32
Cross section of worm wheel.

botics books advocate the use of an ordinary bolt as a worm. This will work provided that the thread of the bolt is carefully matched to the pitch of the spur gear used as the worm wheel.

Internal gears

Instead of the disc shape of a spur gear, an *internal* gear has the shape of an annulus and (as the name implies) teeth on the internal edge of this annulus. Some internal gears will also have teeth on the external edge as shown in Fig. 7-34. Internal gears are often used in planetary gear trains.

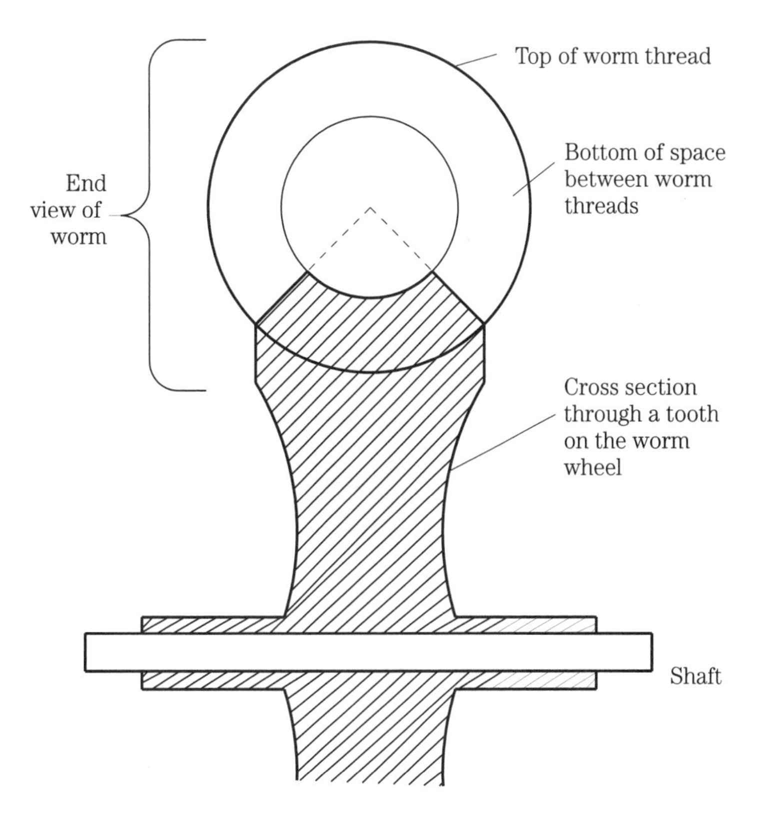

7-33 Worm meshed with worm wheel.

7-34
An internal gear with both internal and external teeth.

Gear sources

Spur gears are very common and they can be found in many household appliances such as electric hand tools, tape players, electric mixers, blenders, and food processors.

Erector sets

The old **Erector** sets made by A. C. Gilbert used to include several different sizes of cast metal spur gears that fastened to shafts using set screws. The super deluxe set even had bevel gears. The new **Erector** sets made in France by Meccano S. A. contain only two sizes of spur gear and one small-lead-angle worm that mates satisfactorily although not ideally with the spur gears. All three gears, which are shown in Fig. 7-35, are plastic molded around brass hubs.

7-35 Gears from Erector set by Meccano S. A.

LEGO Technics

The **LEGO Technics** line of construction toys includes three different sizes of spur gears, the hybrid spur-crown discussed previously, an internal gear-ring, and a worm that meshes with any of the external gears. All of the gears (which are shown in Fig. 7-36) are plastic, and each external gear has a cross-shaped hole that is intended to provide a friction fit over the cross-shaped plastic axles that are part of the **Technics** system. There are also rack sections that can be mounted on the top (peg side) of standard **LEGO** components. Each of the four gears will mate with this rack, but the worm's thread spacing is just a little too close for smooth operation against the rack. The **LEGO** gears mesh almost perfectly with the **Gears-In-Motion** spur-bevel hybrid gears discussed below.

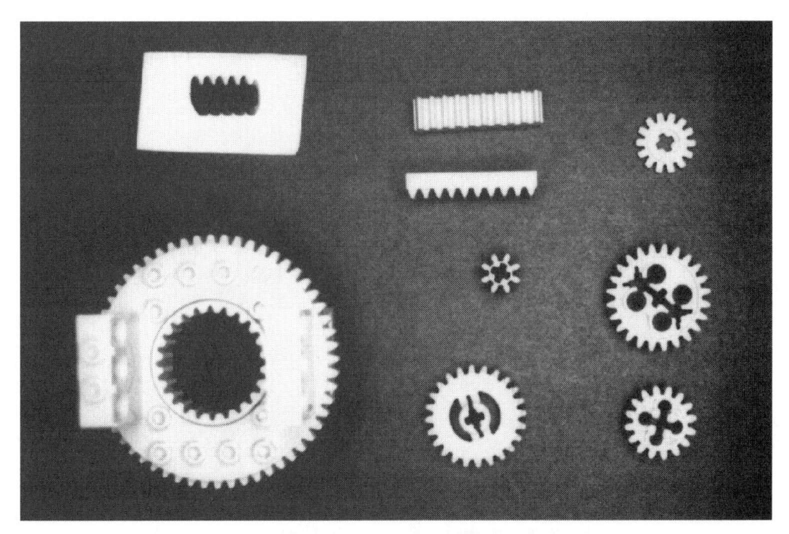

7-36 Gears from the LEGO Technics construction toy.

Gears-In-Motion

Figure 7-37 is a photograph of an unusual gear design that is used in an educational toy called **Gears-In-Motion**. The gear is a spur-bevel hybrid. It is as though a spur gear and bevel gear were pressed together as shown in Fig. 7-38. The set includes this design in three different sizes:

- 2.4-inch outer diameter with 60 teeth
- 1.62-inch outer diameter with 40 teeth
- 0.85-inch outer diameter with 20 teeth

7-37 Spur-bevel hybrid gears from Gears-In-Motion.

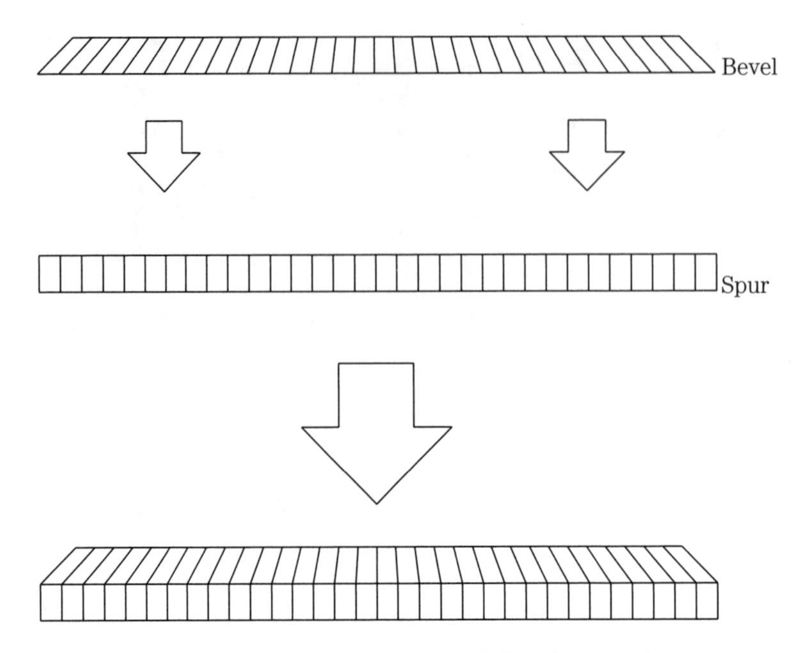

7-38 Gears-In-Motion gear as a composite of a bevel gear and spur gear.

These gears can be used as either spur gears or as 45° bevel gears. In fact, the gears will mesh satisfactorily at any shaft angle between the two limiting cases of spur-gear usage (parallel shafts) and 45° bevel-gear usage (perpendicular shafts). However, the gears should be used at these intermediate angles only in light duty applications because the teeth make only partial contact at these intermediate positions. Actually, the teeth are never in **full** contact. For parallel shafts, the spur portions of the teeth are completely meshed, and for perpendicular shafts, the bevel portions of the teeth are completely meshed. At intermediate angles, the gears mesh over a small part of the tooth profile in the area where the spur portion and bevel portion join together. (See Fig. 7-39.)

Gear trains

A gear train consists of two or more meshed gears. In a *simple* gear train, there is one gear on each shaft. In a *compound* gear train, two or more gears are on a single shaft. In both the simple gear train and the compound gear train, the gears turn on shafts that have fixed positions relative to the frame. Gear trains of this ilk are called *ordinary* gear trains. *Planetary* gear trains, which have some gears on shafts that move relative to the frame, are discussed later in this chapter.

Tooth ratio

Tooth ratio is an important concept in the design of gear trains. For two gears meshed as shown in Fig. 7-40, the tooth ratio is simply the ratio between the num-

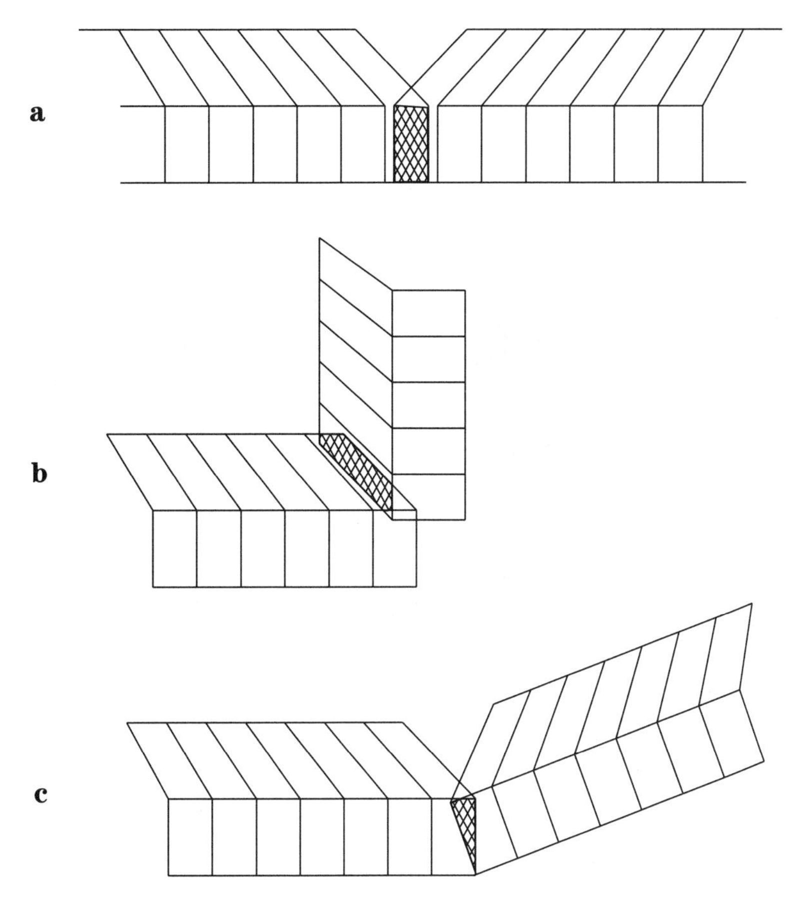

7-39 Various ways that Gears-In-Motion gears can be meshed.

bers of teeth on each gear. As indicated in the figure, gear **A** has N_A teeth and gear **B** has N_B teeth, so we could define the tooth ratio as either N_A**:**N_B or N_B**:**N_A. The speed ratio between two gears is the inverse of the corresponding tooth ratio:

$$\frac{\omega_A}{\omega_B} = \frac{N_B}{N_A} \tag{7-7}$$

The torque ratio between two gears is equal to the corresponding tooth ratio:

$$\frac{\tau_A}{\tau_B} = \frac{N_A}{N_B} \tag{7-8}$$

Example 7-3

For the gears shown in Fig. 7-40, let's assume that $N_A = 19$ and $N_B = 57$. Gear **A** is driven at 600 rpm with a torque of 120 in•oz. Neglecting frictional losses, determine the speed and torque of gear **B**.

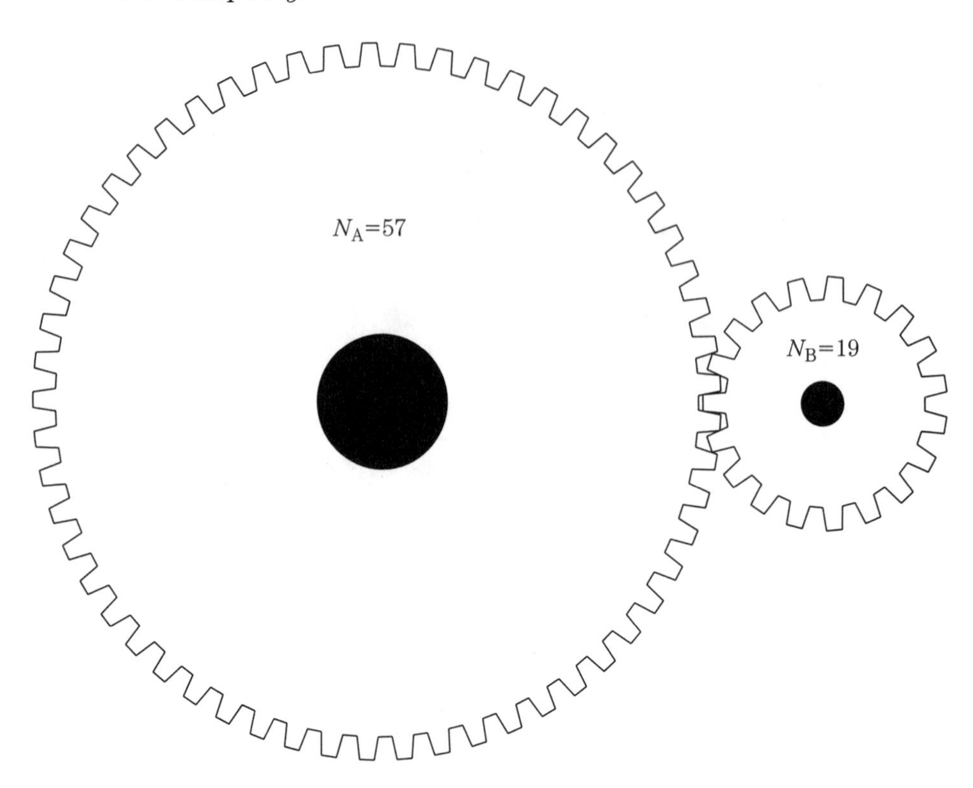

7-40 Gear train for Example 7-1.

Solution: Using Eq. 7-7 for the speed ratio, we obtain:

$$\frac{\omega_A}{\omega_B} = \frac{N_B}{N_A}$$

$$\frac{600}{\omega_B} = \frac{57}{19}$$

$$\omega_B = 200 \text{ rpm}$$

Using Eq. 7-8 for the torque ratio, we obtain:

$$\frac{\tau_A}{\tau_B} = \frac{N_A}{N_B}$$

$$\frac{120}{\tau_B} = \frac{19}{57}$$

$$\tau_B = 360 \text{ in}\cdot\text{oz}$$

The speed and torque relationships between two meshed gears are the same as the speed and torque relationships between two discs in "perfect" (i.e., no slippage) frictional contact where the diameters of the discs equal the pitch diameters of the gears. To save time and effort when designing gear trains, it is customary to depict gears using only their pitch circles plus a notation indicating the number of teeth.

Simple gear trains

A simple gear train is depicted in Fig. 7-41. There is one gear per axis (shaft), and the location of each axis is stationary relative to a fixed frame. The "first" gear in the train, that is, the gear that is turned by the motor or other prime mover, is called the *driving* gear or *input* gear. The final gear that turns the load is called the *driven* gear or *output* gear. The gears between the driving and driven gears are called *idler* gears. In Fig. 7-41, gear **1** is the driving gear, gear **4** is the driven gear, and gears **2** and **3** are idler gears.

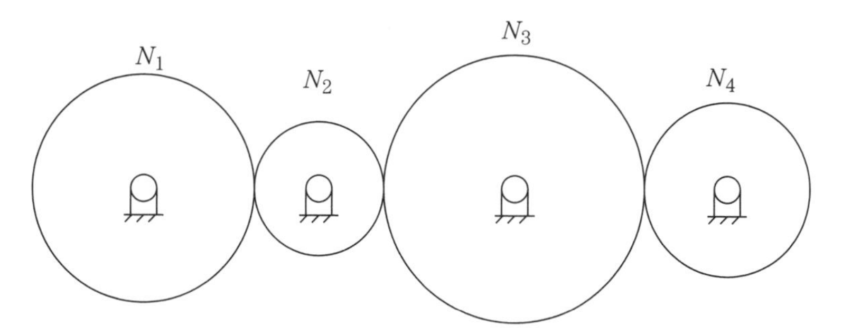

7-41 Simple gear train.

The end-to-end tooth ratio for a simple gear train is simply the product of the individual pairwise tooth ratios:

$$\frac{N_1}{N_2} \times \frac{N_2}{N_3} \times \frac{N_3}{N_4} = \frac{N_1}{N_4}$$

Because the number of teeth for each idler gear appears once in the numerator and once in the denominator, these values always cancel. Therefore, the end-to-end ratio does not depend upon the tooth counts of any of the idler gears. For simple gear trains having an even number of gears, the driving gear and driven gear will rotate in opposite directions. Conversely, for an odd number of gears, the driving gear and driven gear will rotate in the same direction. At first glance, it may seem as though idler gears are of no use other than reversing the direction of rotation. After all, the gear train shown in Fig. 7-41 can be replaced by two gears having a ratio of N_1**:**N_4. Suppose that the shaft positions of gears **1** and **4** must remain in the locations shown in Fig. 7-41. If we wish to eliminate the idler gears when faced with this restriction, we must increase the diameters of gears **1** and **4** as shown in Fig. 7-42 while maintaining a tooth ratio of N_1**:**N_4. Given the shaft-location constraints, the four-gear implementation of Fig. 7-41 has a space saving advantage over the two-gear implementation of Fig. 7-42.

Complicated gear trains may, in fact, be just a composite of several simple gear trains. Consider the case shown in Fig. 7-43. The driving gear is gear **4** and the driven gears are gears **1**, **5**, and **6**. This train is really three simple gear trains that intersect at gears **3** and **4**. For analysis purposes, the train can be separated into the three simple trains shown in Fig. 7-44.

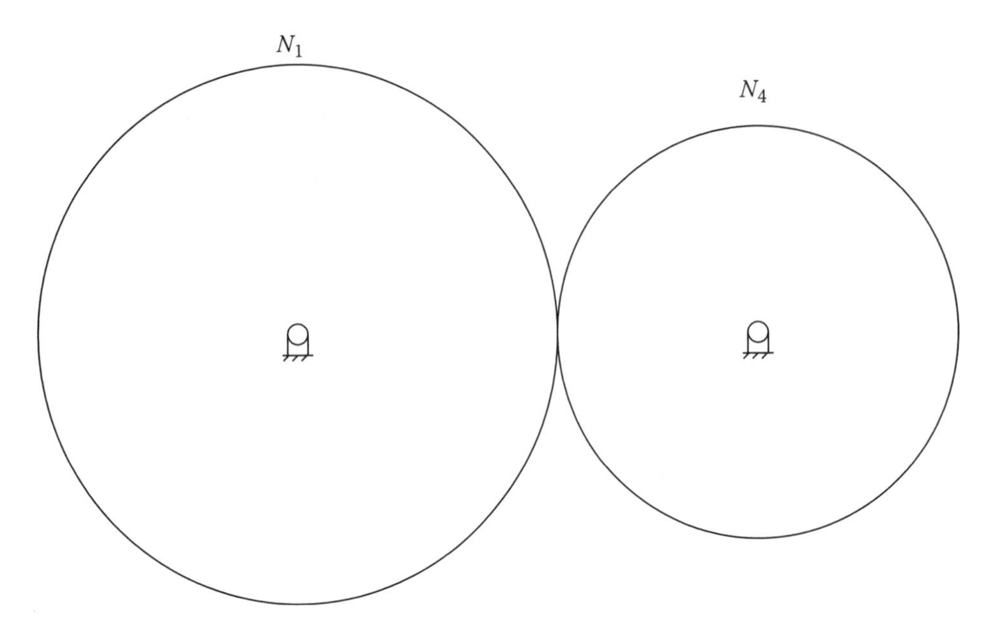

7-42 Gears N_1 and N_4 from Fig. 7-41 made larger to maintain the same center-to-center spacing and the same tooth ratio.

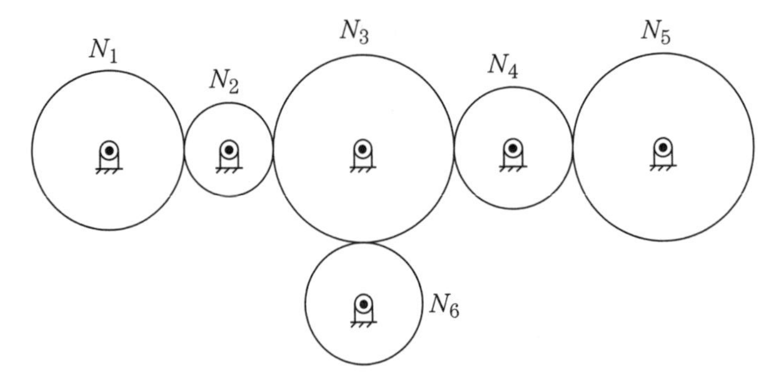

7-43 Complicated gear train. Driving gear is gear 4, and the driven gears are 1, 5, and 6.

Compound gear trains

A compound gear train is depicted in Fig. 7-45. The location of each axis is stationary relative to the fixed frame, and at least one of the axes contains more than one gear. The edge view provided is sometimes necessary to clarify the position of the gears in several different planes. Computation of the end-to-end tooth ratio is just a bit trickier for compound trains than it is for simple trains. For the compound gear train of Fig. 7-45, the pairwise tooth ratios N_1:N_2 and N_3:N_4 can be combined by exploiting the fact that gears 2 and 3 are on the same shaft and rotate at the same speed.

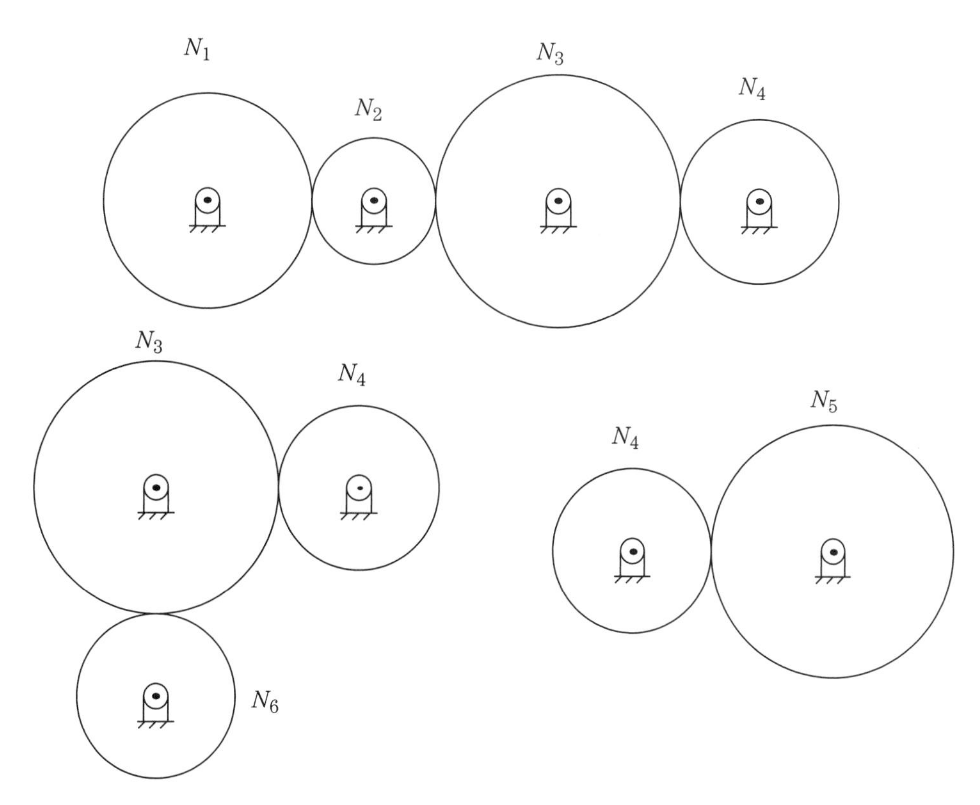

7-44 The simple gear trains comprising the complicated gear trains of Fig. 7-43.

$$\frac{N_1}{N_2} = \frac{\omega_2}{\omega_1} \Leftrightarrow \omega_2 = \frac{N_1}{N_2}\,\omega_1$$

$$\frac{N_3}{N_4} = \frac{\omega_4}{\omega_3} \Leftrightarrow \omega_3 = \frac{N_3}{N_4}\,\omega_4$$

Because $\omega_2 = \omega_3$, we can combine these two equations to obtain:

$$\frac{N_1}{N_2}\,\omega_1 = \frac{N_3}{N_4}\,\omega_4$$

$$\frac{\omega_4}{\omega_1} = \frac{N_1}{N_2} \cdot \frac{N_3}{N_4}$$

Unlike simple gear trains, the end-to-end tooth ratio for compound gear trains **does** depend upon the tooth counts of the idler gears.

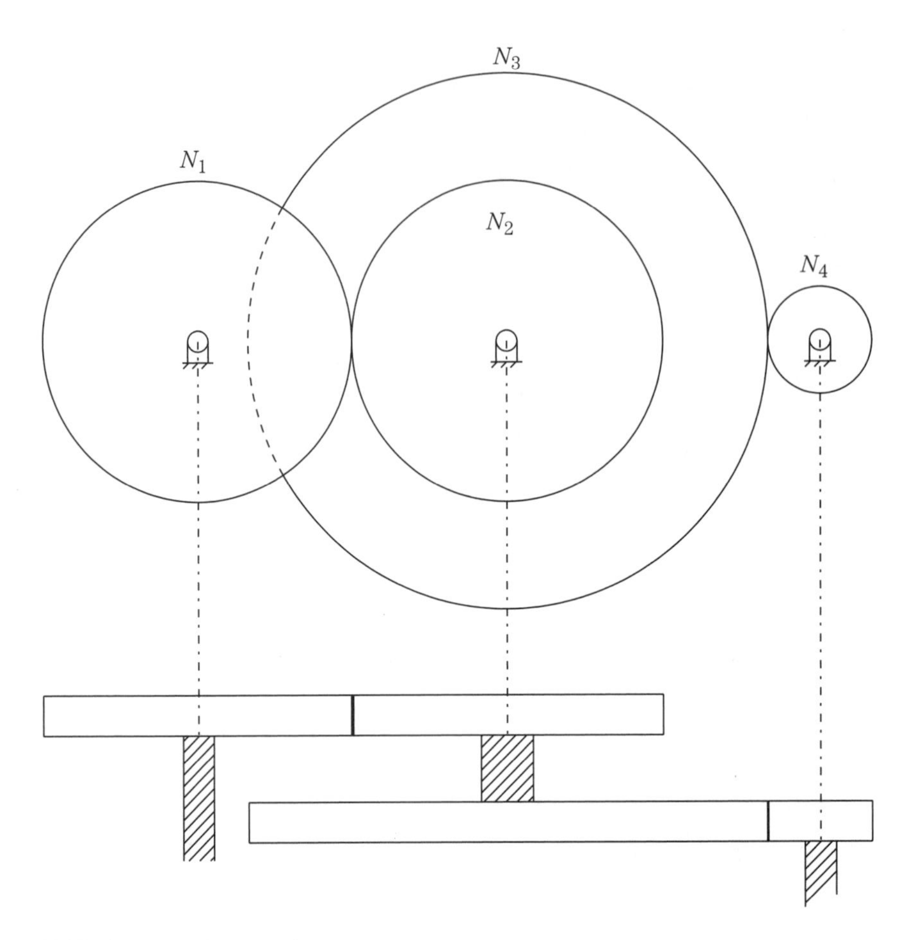

7-45 Compound gear train.

Planetary gear trains

In a *planetary* gear train, some of the gears are on shafts that move relative to the frame. A simple planetary gear train is depicted in Fig. 7-46.

This gear train (as well as most other planetary gear trains) is said to have *two degrees of freedom*. This means that two inputs or constraints must be provided in order to define the behavior of the gear train. Some possible combinations of inputs include:

- Gear **A** rotated around its shaft with gear **B** rotated around its shaft. The arm will rotate around the shaft of gear **B** as needed to accommodate the movement of gear **A** around the circumference of gear **B**.
- Gear **B** is fixed with gear **A** rotated around its shaft. The arm will rotate around the shaft of gear **B** as needed to accommodate the movement of gear **A** around the circumference of gear **B**.
- Gear **B** is fixed with the arm rotated around the shaft of gear **B.** Gear **A** will rotate about its shaft and move around the circumference of gear **B** as needed to accommodate the movement of the arm.

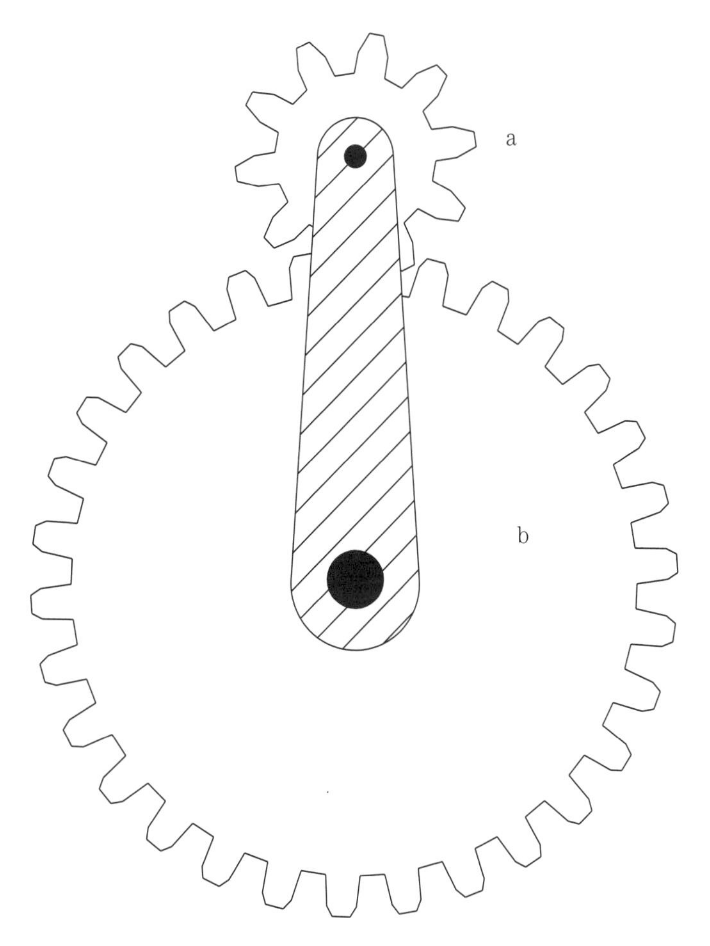

7-46 A simple planetary gear train.

Example 7-4

In the Black and Decker Model 9072 cordless power screwdriver, two planetary gear trains are cascaded as shown in Fig. 7-47. The motor shaft and the 6-tooth pinion attached thereto must make 9 complete revolutions relative to the housing in order for the first planetary carrier and its attached 6-tooth pinion to make one complete revolution relative to the housing.

Likewise, this 6-tooth pinion must make 9 complete revolutions relative to the housing in order for the second planetary carrier and the attached output shaft to make one complete revolution relative to the housing. Therefore, we can say that for each revolution of the output shaft, the motor shaft must make 81 complete revolutions. The no-load output speed of the shaft is 130 rpm, so the motor is turning at 10,530 rpm. With a mechanical advantage of 81, this tool can use a relatively small battery-powered motor and yet still deliver considerable torque at the output shaft.

7-47 The planetary gear train contained in the Black and Decker Model 9072 cordless power screwdriver.

8
CHAPTER
Other mechanical stuff

This chapter is a collection of useful information on miscellaneous mechanical items that don't fit into any of the other chapters.

Shafting

Gears, sprockets, pulleys, and wheels for friction drives all share a common need; each must usually be placed on some sort of shaft or axle in order to be used. This requirement can really be separated into two similar but distinct requirements. Sometimes a wheel must turn on a shaft and sometimes the wheel must be attached so that it turns with the shaft. Still other times, a ball bearing will be placed between the wheel and the shaft with the inner diameter of the inner race being a tight fit over the shaft, and the outer diameter of the bearing fitting tightly in the wheel.

Figure 8-1 shows a wheel that turns directly on a shaft. This will sometimes be adequate provided that the center hole in the wheel is smooth and a good fit (not too loose, not too tight) for the axle. For very light-duty applications, the shaft may not even need to be cylindrical. As long as the shaft doesn't bind or wobble in the hole, cruciform or hexagonal shafts can often be used.

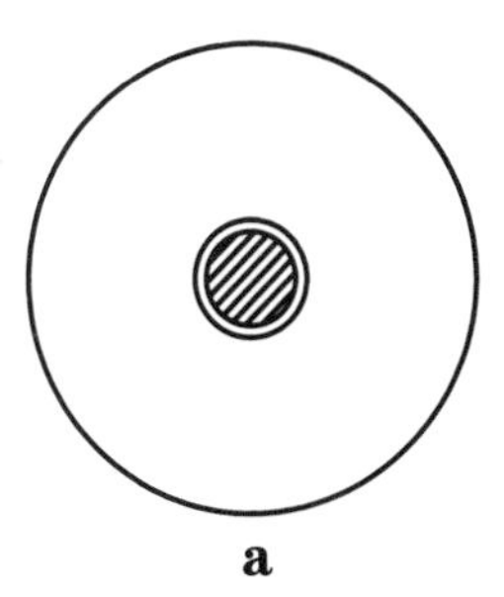

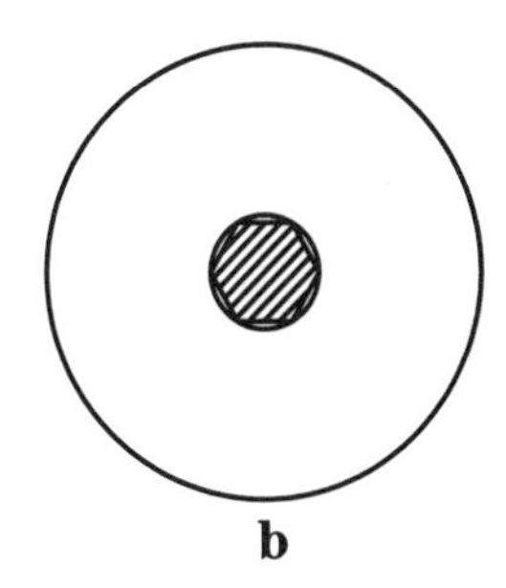

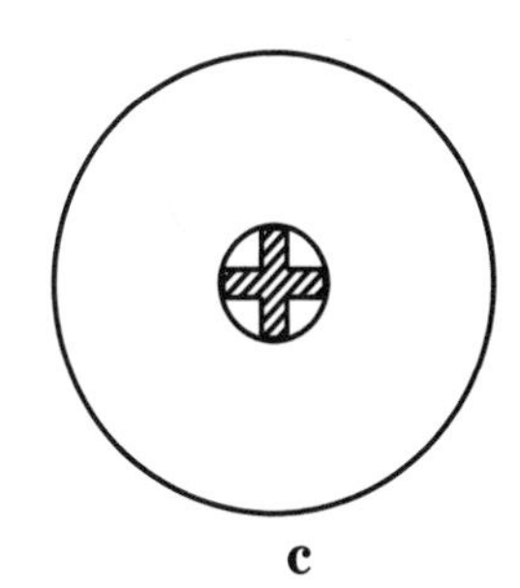

8-1 Wheels that turn directly on a shaft: (a) round shaft slightly smaller than hole in wheel, (b) hexagonal shaft, (c) cruciform shaft.

Shaft sizing

As depicted in Fig. 8-2, a shaft can be subject to two different types of loading: torsional and bending. The maximum combined torsional and bending load L that a shaft should be subjected to is given by:

$$L = s_{max} z_p \tag{8-1}$$

where s_{max} is the maximum allowable shear stress (in psi) for the shaft material and z_p is a geometry-dependent quantity called the *polar section modulus*. Values of s_{max} for various materials are listed in Table 8-1, and formulas for computing z_p for a variety of shaft cross sections are presented in Fig. 8-3.

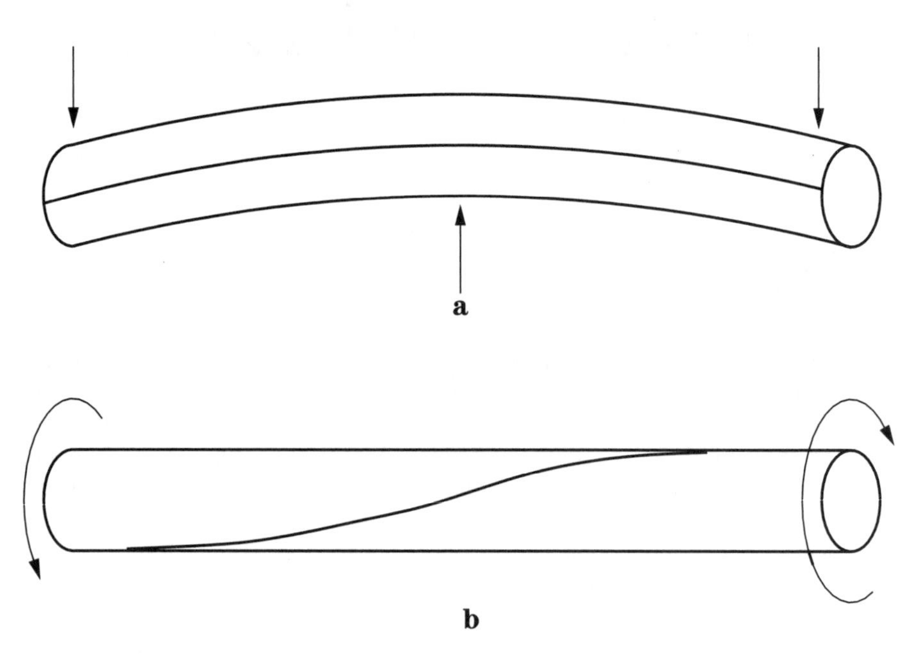

8-2 Two types of shaft loading: (a) bending load, (b) torsional load.

Material	s_{max}
stainless steel	12,000
steel	8000
aluminum	5000
brass	4000
polycarbonate	1600
plexiglas	1400
hardwood doweling	200

Section	z_p

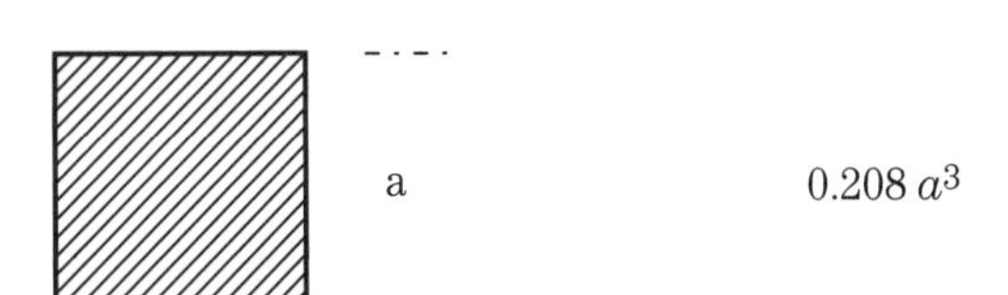

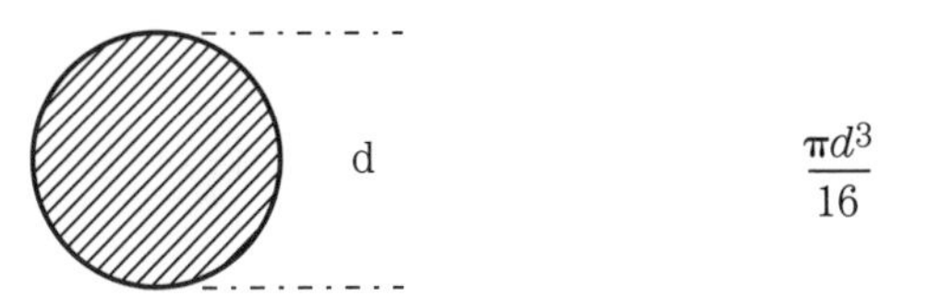

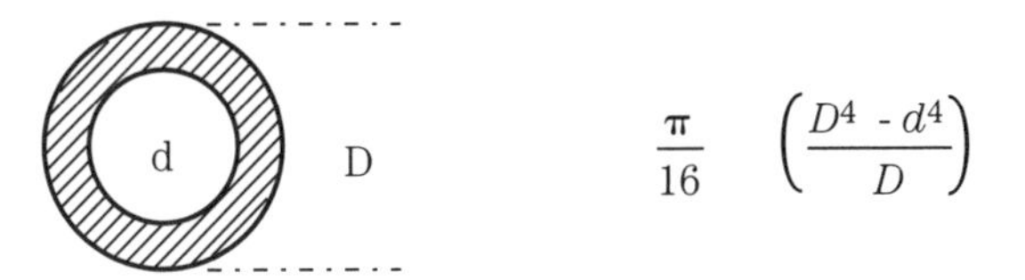

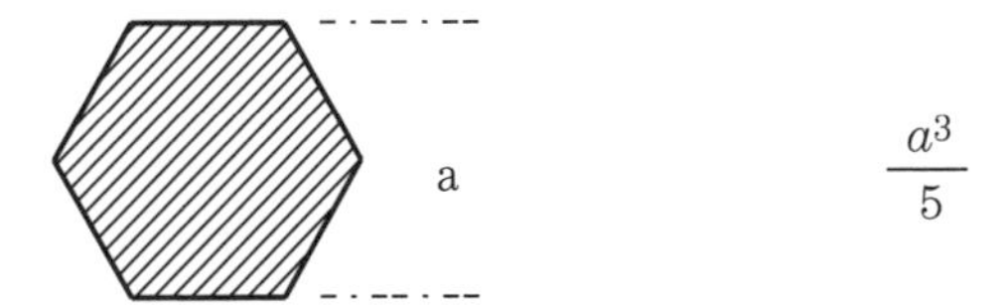

8-3 Formulas for computing polar section modulus z_p.

For the usual case of solid cylindrical shafting we find:

$$z_p = \frac{\pi d^3}{16} \tag{8-2}$$

where d is the cross-sectional diameter. Combining Eq. 8-1 and Eq. 8-2 we obtain:

$$L_{max} = \frac{s_{max} \pi d^3}{16} \tag{8-3}$$

This formula can be used to determine the maximum allowable load for a given shaft.

Example 8-1

Determine the maximum allowable load for a ½ inch plexiglas shaft.

Solution:

From Table 8-1, s_{max} = 1400 psi for plexiglas. Thus we obtain

$$L_{max} = \frac{(1400)\pi(0.5)^3}{16}$$

$$= 34.36 \text{ in} \bullet \text{lb}$$

Equation 8-3 can be "turned inside out" to yield the minimum diameter d_{min} needed for a load of L:

$$d_{min} = \sqrt[3]{\frac{16L}{s_{max}\ \pi}}$$

Example 8-2

Determine the minimum diameter for a cylindrical stainless steel shaft given the load of 34.36 in•lb obtained in Example 8-1.

Solution:

From Table 8-1, s_{max} = 12,000 psi for stainless steel. Thus we obtain

$$d_{min} = \sqrt[3]{\frac{(16)(34.36)}{\pi(12000)}}$$

$$= 0.2443$$

Notice that because the diameter is inversely proportional to the cube root of s_{max}, relatively large changes in s_{max} make small changes in d_{min}. For an almost eight-fold increase in s_{max} from 1400 to 12,000, the value of d_{min} is only halved, from 0.5 to 0.2443.

So far we have assumed a value for the maximum allowable combined loading L_{max}. Just where does this value come from? For the case of minor shocks in bending load and moderately sudden applied torque, the combined load can be obtained as:

$$L = 2\sqrt{M^2 + \tau^2}$$

where

M = bending moment in•lb

τ = torque in•lb

Bearings

Bearings are used to support axles, shafts, and other items that must be constrained in some directions and allowed to move in others. In addition to bearings that can be purchased, it is possible to fabricate some types of bearings in the home shop. Bearings can be classified in two general groups: *plain* bearings, and *rolling-contact* bearings.

Plain bearings

Plain bearings are characterized by sliding contact between surfaces. Many common household items and toys contain plain bearings. They are simple and inexpensive for light-duty applications. However, with the appropriate lubrication system they can be used in fairly heavy-duty applications. The main bearings in automobile engines are plain bearings. For many light-duty applications, no lubrication **system** is required at all; periodic application of a few drops of grease or oil will often suffice.

It would be a bad idea to have a rotating steel axle pass directly through a hole in an aluminum chassis member. Even if frictional losses and wear are not concerns, the loud screeching noises will be. A simple design fix is shown in Fig. 8-4. The heavy plastic cutting boards sold in kitchen supply stores are a good source of plastic for bearing blocks. These boards are made from one of two different kinds of plastic. Stay away from the clear, rock-hard acrylic boards. The softer, almost slippery-feeling white boards are made from a plastic that is almost ideal for light-duty bearings.

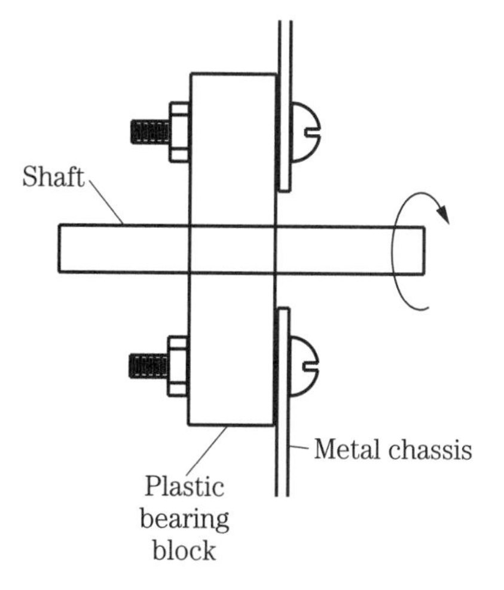

8-4
A plastic bearing block used to avoid metal-to-metal contact between rotating shaft and metal chassis.

Ball bearings

An exploded view of a typical ball bearing is shown in Fig. 8-5. The four major components are the outer race, inner race, separator, and balls. The outer race is typically press-fit into a hole in some sort of frame or maybe a hole in the center of a wheel or pulley. The inner race is press-fit over an axle or shaft. The inner race and outer race (along with the items attached to them) are free to rotate relative to each other by riding on the balls which are between the races.

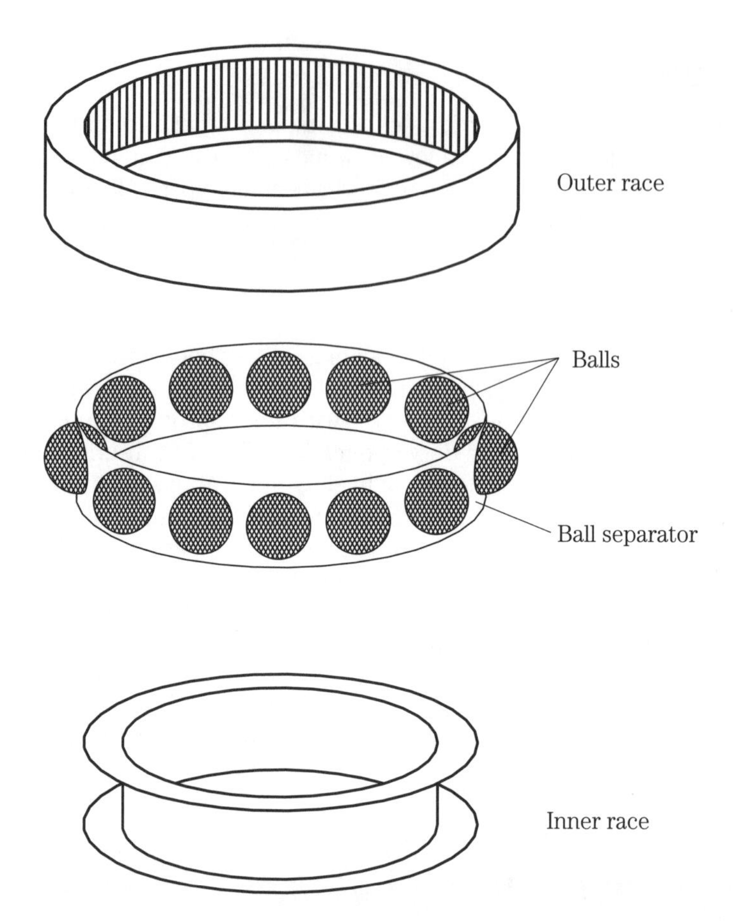

8-5 Exploded view of a typical ball bearing.

A separator keeps the balls evenly spaced around the circumference of the inner race. Various small sizes of ball bearings are available from hobby shops that cater to the R/C racing market. A range of slightly larger ball bearings can be found at the local purveyor of customized rollerskates and skateboards.

One peculiar sort of ball bearing may be of particular interest in the robotics area. Rubbermaid and a few other companies make rotating plastic turntables that are sometimes called "lazy Susans." These turntables are sold for use in kitchen cabinets with the idea that instead of digging through the rubble on the bottom of the cabinet, you carefully organize everything on the top of the turntable. This allows you to find things easily by spinning the table so that everything passes by the front of the cabinet where it can be clearly seen. A disassembled turntable is shown in Fig. 8-6. Notice the large, albeit sparsely balled, ball bearing. The turntable with this bearing included can be modified for applications where a large bearing is needed for a light load. An example that springs to mind is a swiveling neck joint between a robot's head and body.

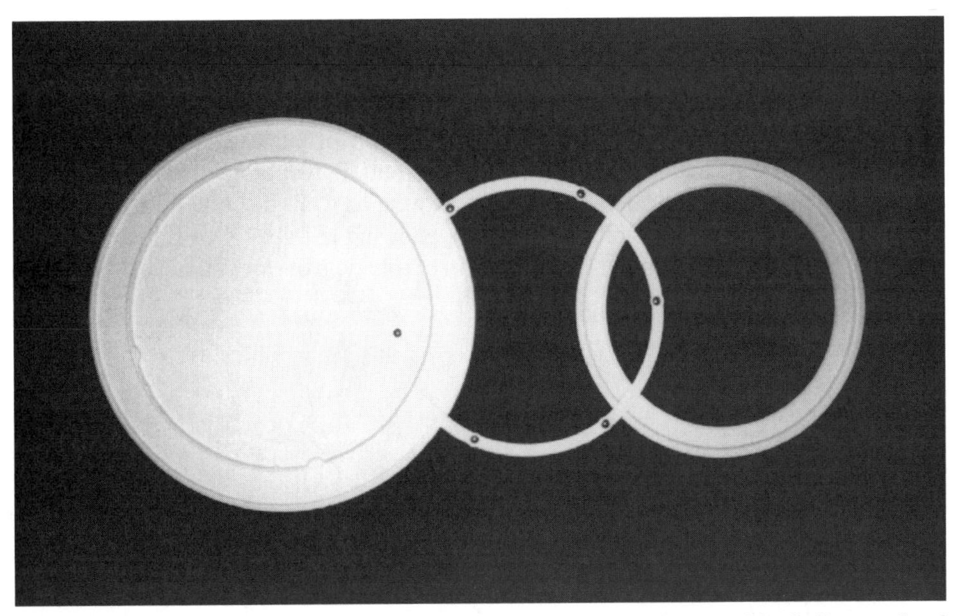

8-6 Disassembled "lazy susan" turntable. Left to right: movable turntable, ball race, fixed baseplate.

Ratchets

From a distance, ratchets resemble gears. However, up close, ratchets and gears are very different. A gear and a ratchet are compared in Fig. 8-7. As discussed in chapter 7, gear teeth are shaped to most effectively transfer motion and force to the teeth of a mating gear. The preferred tooth shape is a curve—usually an involute, sometimes a cycloid. On the other hand, ratchet teeth profiles are often composed of straight lines or arbitrary curves and resemble the teeth of a circular saw blade.

8-7 Comparison between a gear (a) and a ratchet (b).

Basic principles

Ratchets can be used for several different purposes. One use is as a sort of "directional clutch" to transmit rotary motion in one direction but not the other. A ratchet drive for a socket wrench is a familiar example of this use. With the direction control set one way, a clockwise rotation of the handle will be transmitted through the ratchet and cause the socket to turn clockwise as well. If the handle is then rotated counterclockwise, the ratchet will slip, allowing the socket to remain stationary. The basic principles behind the operation of this type of ratchet can be explained using Fig. 8-8.

8-8
A simple ratchet.

As the lever arm (a) is moved in a counterclockwise direction around the pivot (b), the end of pawl (d) engages against one of the steep tooth faces on ratchet (c) and causes the ratchet to turn in a counterclockwise direction. If the lever arm is then moved in a clockwise direction around pivot (b), the pawl (d) [which is free to swing around on pivot (e)] will "ride up" on the sloping tooth faces of the ratchet and thus allow the ratchet to remain stationary. If the lever (a) is rapidly moved back and forth, an intermittent counterclockwise rotation will be imparted to the ratchet.

Reversible ratchets

Figure 8-9 illustrates the basic principles of a ratchet for operation in one direction, but the ratchet drive for a socket wrench set must be capable of selectable operation in two directions: clockwise to tighten bolts (assuming right-hand threads) and counterclockwise to loosen bolts. The ratchet shown in Fig. 8-8 is inherently unidirectional because of the nonsymmetrical shape of its teeth. Reversible ratchets, like the one shown in Fig. 8-9, have symmetric (and usually steep-sided) tooth profiles, with the ratchet action being provided by the shape of the pawl.

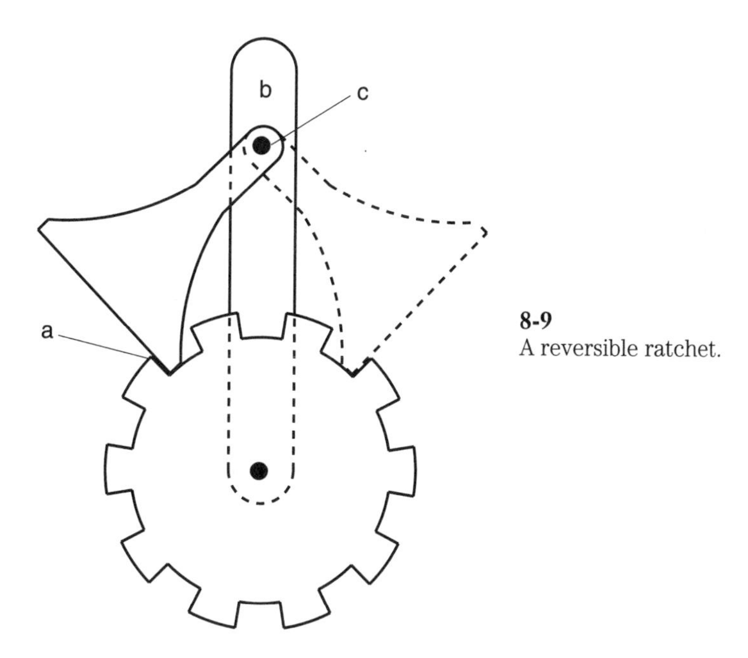

8-9
A reversible ratchet.

When the pawl is in the position indicated by the solid line and the handle (b) is moving in a counterclockwise direction, the steep side of the pawl pushes against the tooth at (a) and the ratchet moves in a counterclockwise direction. If the handle is moving in a clockwise direction, the curved side of the pawl rides up over the ratchet teeth, thus allowing the ratchet to remain stationary. The direction of the ratcheting operation is reversed by flipping the pawl around on pivot (c) to the position indicated by the dotted line.

Springs

Springs come in a wide variety of types, strengths, and sizes. Sometimes selecting just the right spring can make the difference between a mechanism that works smoothly and reliably and a mechanism that barely works at all. The types of springs likely to be useful to experimenters include: helical compression springs, helical extension springs, helical torsion springs, flat springs, and power springs.

Helical compression springs

What do retractable ballpoint pens and the intake valves on automobile engines have in common? Answer: They both make use of *helical compression springs*. The pen spring is one of the weakest likely to be found in an average household, and the valve spring is probably one of the strongest. Helical compression springs are coil springs that "squash" or compress during normal operation. A certain amount of force is required to compress the spring, and when this force is removed the spring will return to its original length. An assortment of helical compression springs is shown in Fig. 8-10.

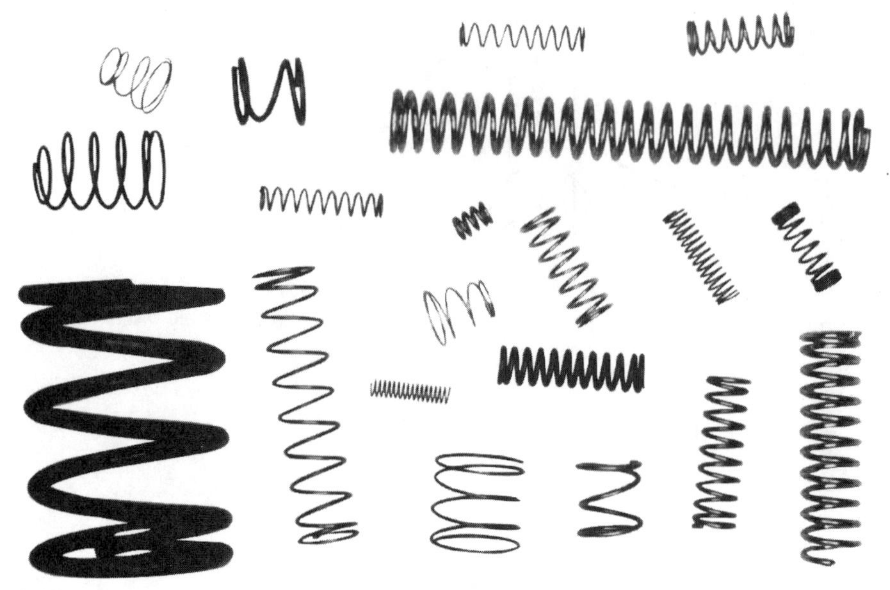

8-10 Assorted helical compression springs.

Some of the important dimensions for a helical compression spring are shown in Fig. 8-11. The *free length* is the overall length of the spring without any loading. The *solid height* is the length of the spring when it is loaded with enough force to close all the coils. If the coils of a spring go in the same direction as the threads of a "normal" bolt, the spring has a *right-hand helix*. If the coils go the other way, the spring has a *left-hand helix*. (See Fig. 8-12.) If two springs are nested one inside the other, they should have opposite helix directions to avoid tangling. If a spring is placed over a threaded bolt or machine screw, the direction of the spring's helix should be opposite to the direction of the threads.

If a spring is too long relative to its diameter (free length more than 4 times the mean coil diameter), it will have a tendency to buckle under compression as shown in Fig. 8-13. Buckling can be avoided in some designs by placing the spring over a guide rod or inside a guide cavity as shown in Fig. 8-14.

The inside diameter is of particular interest if the spring will be placed over a guide rod, and the outside diameter is of interest if the spring will be placed inside a cavity. Under compression, the O.D. of the spring will increase, and the I.D. of the cavity must be large enough to accommodate the increased diameter. If the spring is compressed to its solid height, the increased coil diameter is given by:

$$\mathrm{OD}_{\mathrm{solid}} = \sqrt{D^2 + \frac{p^2 - d^2}{\pi^2}} + d$$

where

D = mean coil diameter
d = diameter of the spring wire
p = pitch

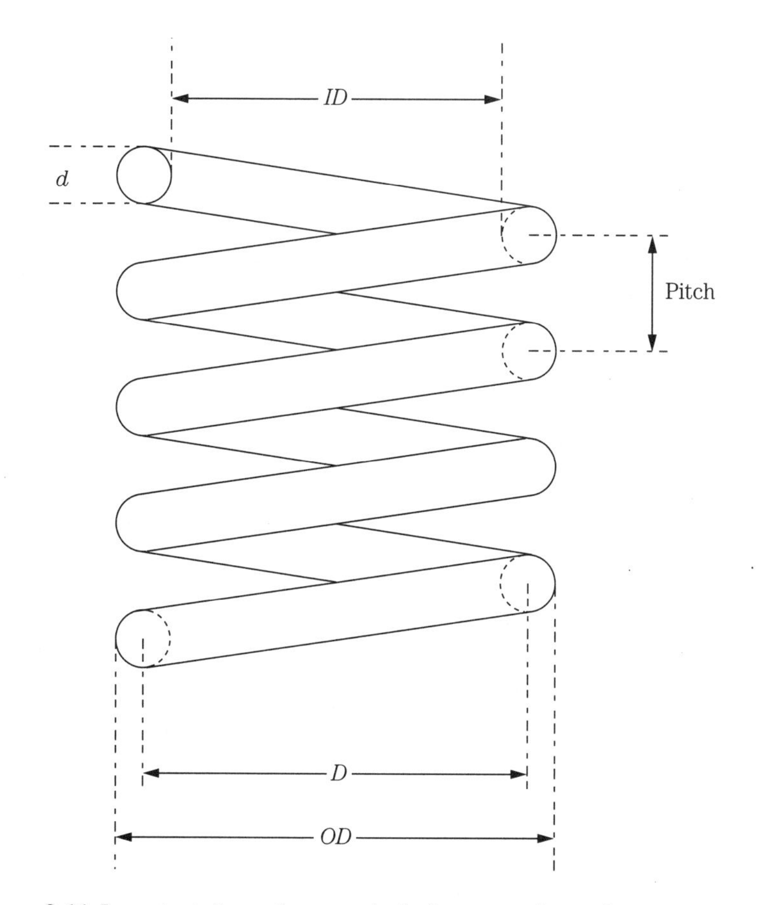

8-11 Important dimensions on a helical compression spring.

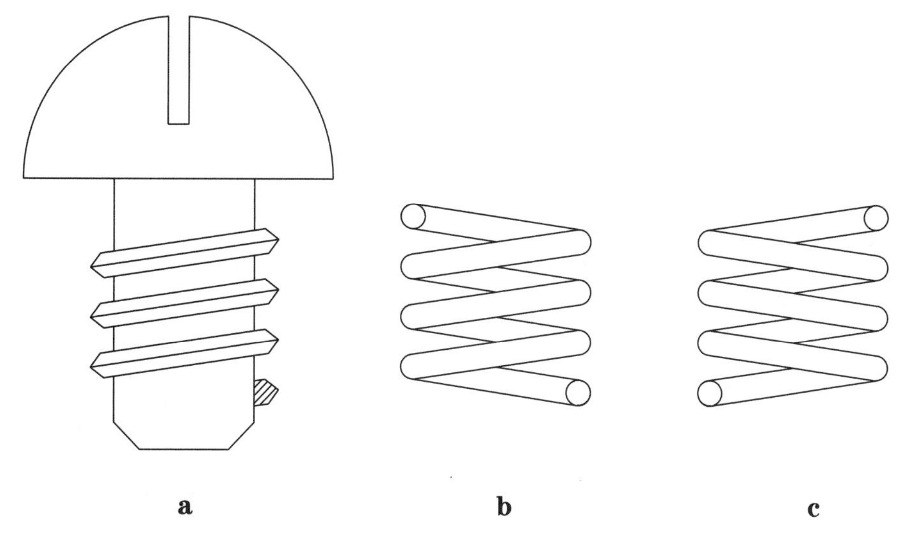

8-12 Helix directions: (a) bolt with "normal" right-hand thread, (b) right-hand coil spring, (c) left-hand coil spring.

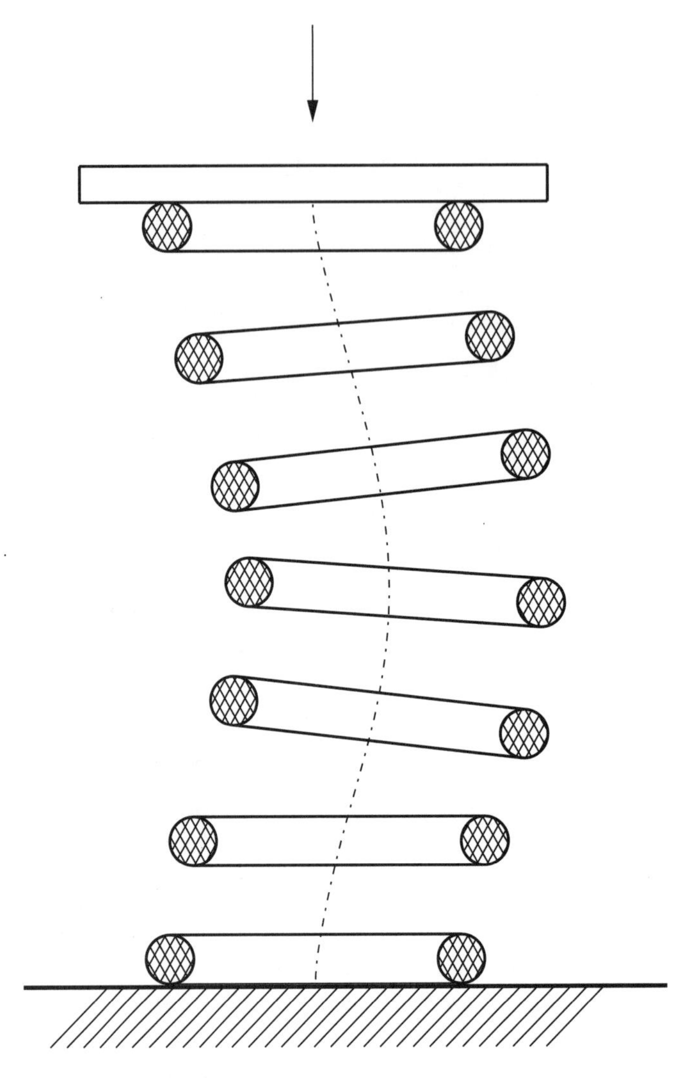

8-13 A spring buckling under compression.

Helical extension springs

The helical compression springs discussed above start out with space between the coils, which closes up as a compressive load is applied to the ends of the spring. In *helical extension springs* the opposite is true; the coils start out "closed" (adjacent coils touching) and space between the coils begins to open up as a tension load is applied to the ends of the spring. In most cases the spring will have a certain amount of *initial tension* which holds the coils together and which must be overcome before coil separation can begin. An assortment of helical extension springs is shown in Fig. 8-15. Most of these springs have loops or hooks at the ends for attaching the springs to the devices that will use them, but some of the springs are supplied with long straight tails that can be cut and bent to the desired hook shape.

8-14 Designs to avoid spring buckling: (a) spring over guide rod, (b) spring inside guide cavity.

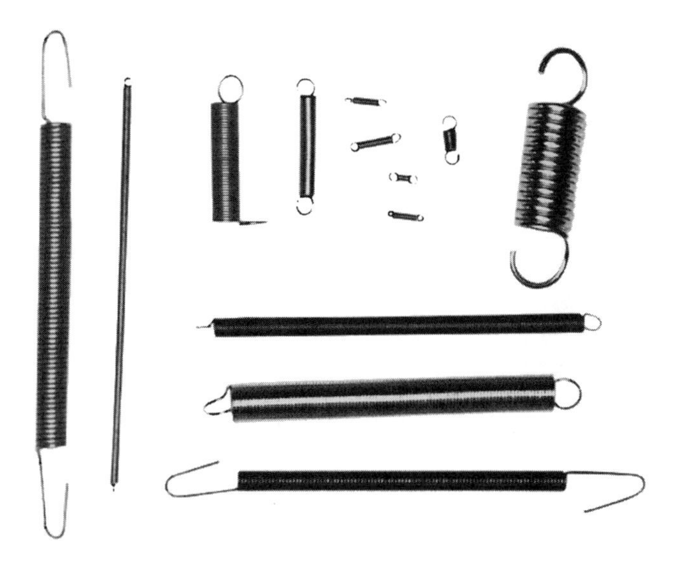

8-15 Assorted helical extension springs.

Helical torsion springs

Helical springs that are designed to exert a torque are called *helical torsion springs*. An assortment of helical torsion springs is shown in Fig. 8-16. The usual practice is to place the coils over a shaft and fix one end as shown in Fig. 8-17. The other end is loaded in the direction that tends to tighten the coils.

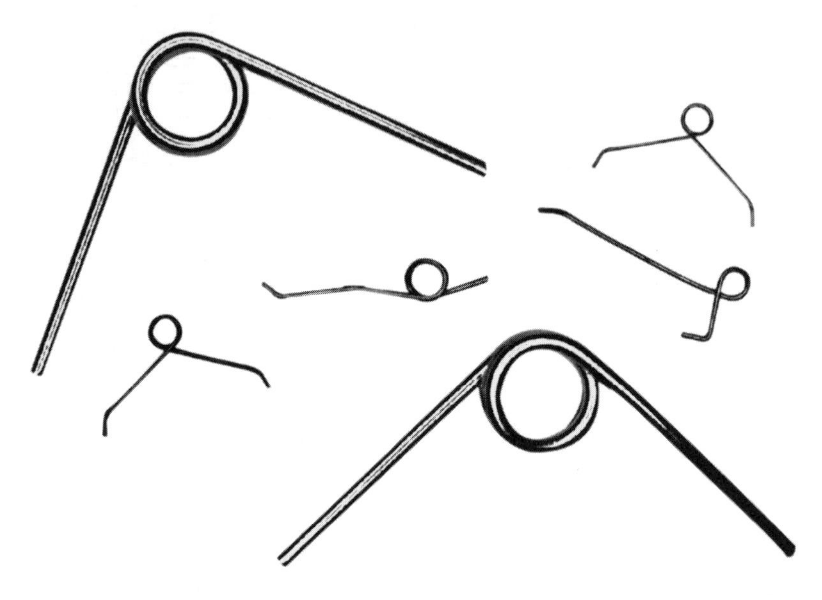

8-16 Assorted helical torsion springs.

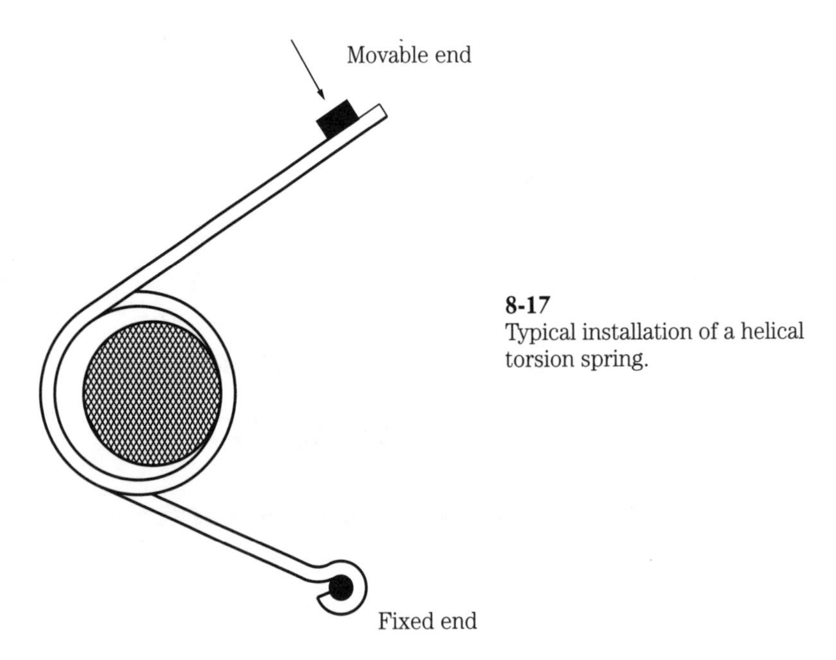

8-17
Typical installation of a helical torsion spring.

The end could be loaded in the direction that tends to unwind or enlarge the coils, but this is not recommended. Loading in the recommended direction increases the length of the spring body and decreases the diameter of the body coils. The length of the loaded spring is given by:

$$L = d(N + 1 + \theta)$$

where

d = diameter of the spring wire
N = number of coils in relaxed spring
θ = deflection in revolutions

Note that the deflection is in units of revolutions, i.e., a deflection of 45 degrees would be expressed as 0.125 revolutions. The diameter of the loaded spring is given by:

$$D = \frac{D_I N}{N+\theta} \tag{8-4}$$

where

D_I = mean coil diameter of relaxed spring
N = number of coils in relaxed spring
θ = deflection in revolutions

The shaft over which the spring is placed should have a diameter equal to about 90 percent of the smallest inner diameter expected to be produced by deflection during operational loading. The smallest inner diameter is obtained by subtracting one wire diameter d from the smallest mean diameter value produced by Eq. 8-4.

Kinematics

Kinematics is concerned with the motions of various parts of a *mechanism* or *linkage* without particular regard to the forces that produce these motions. In kinematic analysis a linkage is composed of rigid members called *links* and the joints that connect the various links together. These joints are sometimes called *kinematic pairs*.

Planar linkages

A typical planar linkage is shown in Fig. 8-18. This linkage is composed of flat strips of metal, wood, or plastic joined together by pins or bolts at the four points shown. Each of the strips is free to rotate around the joint while remaining in a plane that is perpendicular to the axis of the joint pin. Sometimes the entire linkage is loosely spoken of as lying in a plane (hence the name **planar** linkage), but this would be strictly true only if the various links had zero thickness.

In a real linkage of the sort shown in Fig. 8-18, each strip will have nonzero thickness and hence all of the strips cannot lie precisely in the same plane. However, since all of the motion is confined to several parallel planes, the motion of the linkage can be analyzed **as though** it were all confined to the same plane. There are three fundamentally different kinds of joints in planar linkages. The joints of the type shown in Fig. 8-18 are called *revolute* joints. A second type of joint is the *sliding* or *pris-*

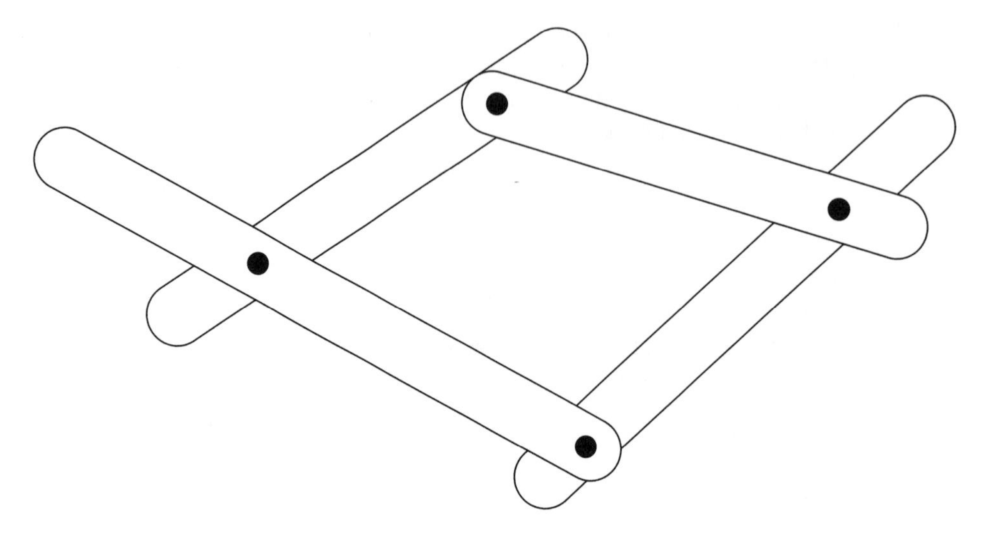

8-18 A typical four-bar linkage.

matic joint shown in Fig. 8-19. A third type of joint, shown in Fig. 8-20, combines both revolute and sliding motions.

The disposition of a revolute joint can be specified in terms of the angle between the two links as shown in Fig. 8-21. On the other hand, the disposition of a prismatic joint can be specified in terms of a distance between the slider and some fixed reference point on the other link. For a joint that both rotates and slides, a distance and an angle are needed to completely specify the disposition of the joint as shown in Fig. 8-22.

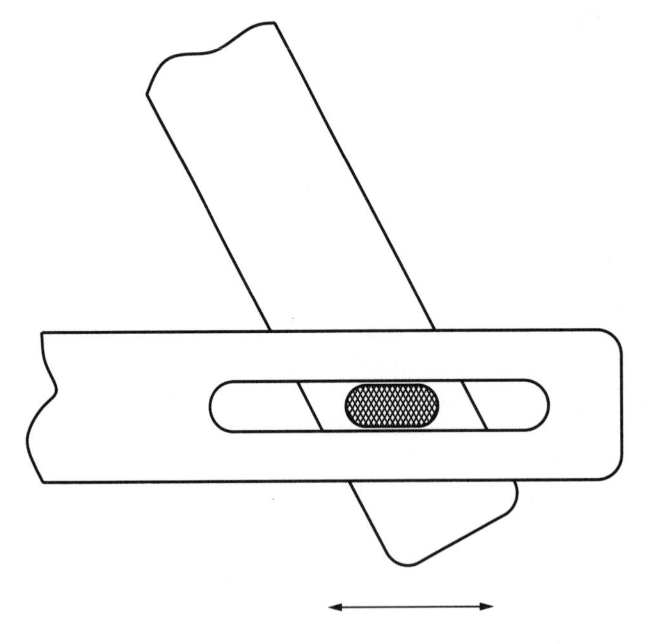

8-19 A sliding or prismatic joint.

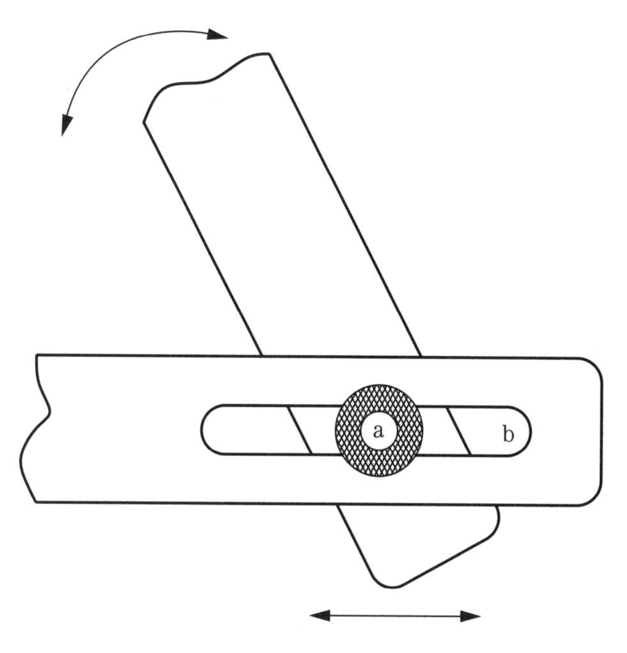

8-20 A joint that both slides and rotates.

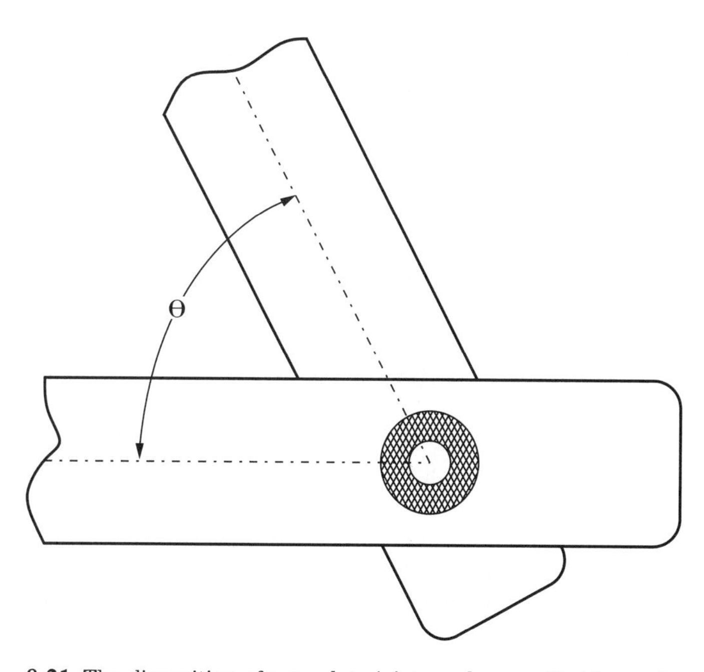

8-21 The disposition of a revolute joint can be specified by angle θ between the two links.

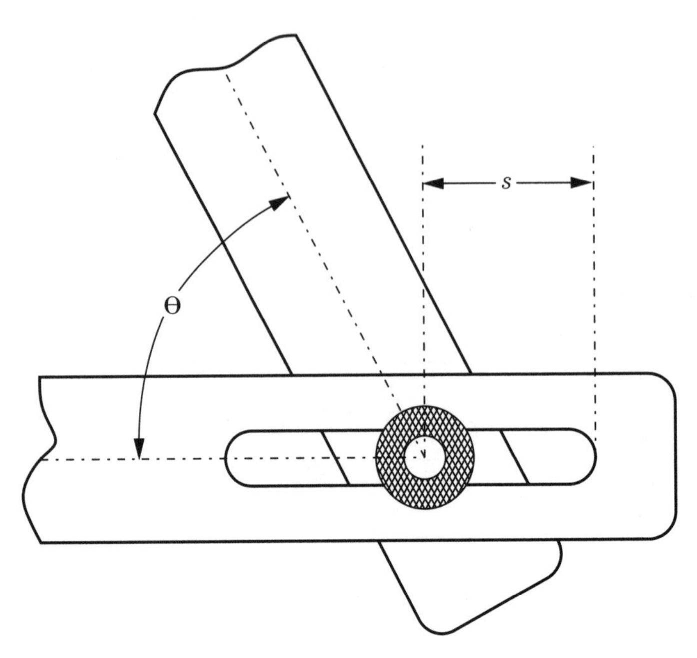

8-22 The disposition of a joint that both rotates and slides is specified in terms of a distance *s* and an angle θ.

Four-bar linkage

An example of the "classic" four-bar linkage is shown in Fig. 8-23. When analyzing the kinematics of a linkage, it is usually more convenient to work with the schematic form shown in Fig. 8-23(b) rather than the pictorial form of Fig. 8-23(a). Usually, one of the links is considered as fixed. For purposes of this discussion, let's assume that link **d** is the fixed link. To emphasize the fact that one link is considered fixed, the four-bar linkage is often drawn as in Fig. 8-23(c). The hash marks below link **d** are a standard method used in mechanism drawings to show that something is fixed. Each of the two links connected to the fixed link is called a *crank* if it can rotate a complete 360 degrees around the pivot connecting it to the fixed link. If the motion of one of these links is restricted to something less than 360 degrees, the link is called a *rocker*. The link that is opposite to (and not directly connected to) the fixed link is called the *coupler*.

The linkage depicted in Fig. 8-23 is somewhat of an abstraction, and it may not at first seem to be of much practical use. The abstraction is for ease of analysis. To convince ourselves that four-bar linkages are indeed useful, let's look at one possible implementation. The leg mechanism for an insect robot is shown in Fig. 8-24. The fixed link extends between pivots **A** and **D**. The rocker between points **D** and **C** is fastened to gear **E**, which is driven by gear **F**. As we will see in the sections below, the relative motions of gear **E** and foot **G** depend upon the relative lengths of the links.

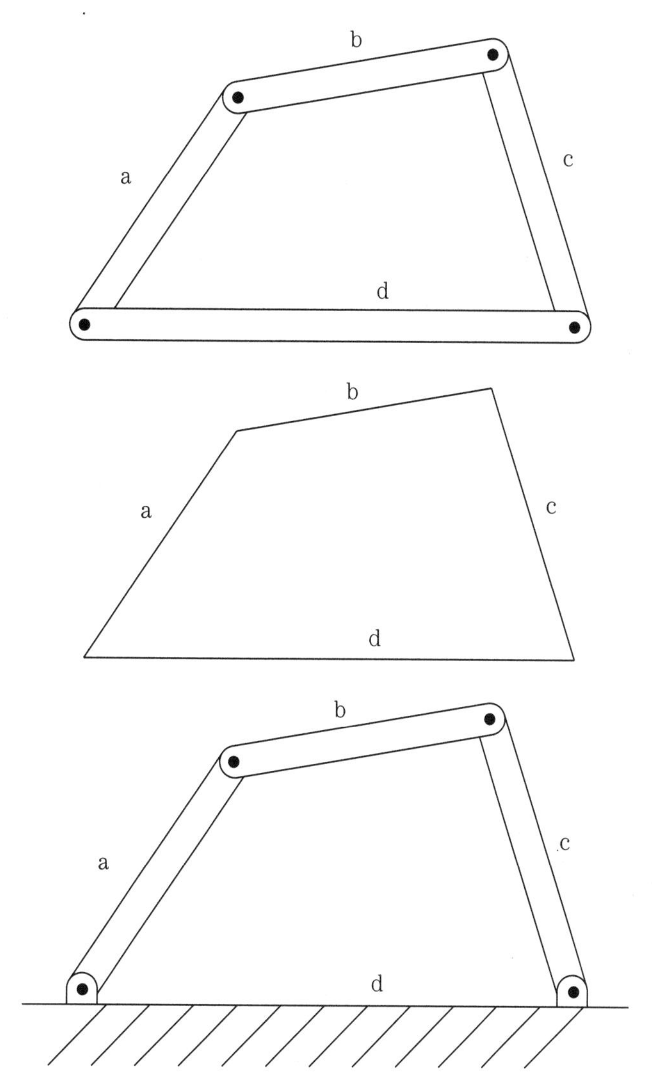

8-23 Three ways of depicting the "classic" four-bar linkage.

Types of four-bar mechanisms

Consider the four-bar linkage shown in Fig. 8-23. Let **a** denote the shortest link with length a. Similarly, let **b** denote the longest link with length b. The remaining two links, **c** and **d**, have lengths c and d respectively. Let's assume that the sum of lengths $a+b$ is less than $c+d$. (This condition with the sum of the shortest and the longest links being less than the sum of the other two links is called the *Grashoff rule*.) Three different *inversions* of this mechanism shown in Fig. 8-25 are the *drag-link* mechanism, *crank-rocker* mechanism, and *double-rocker* mechanism.

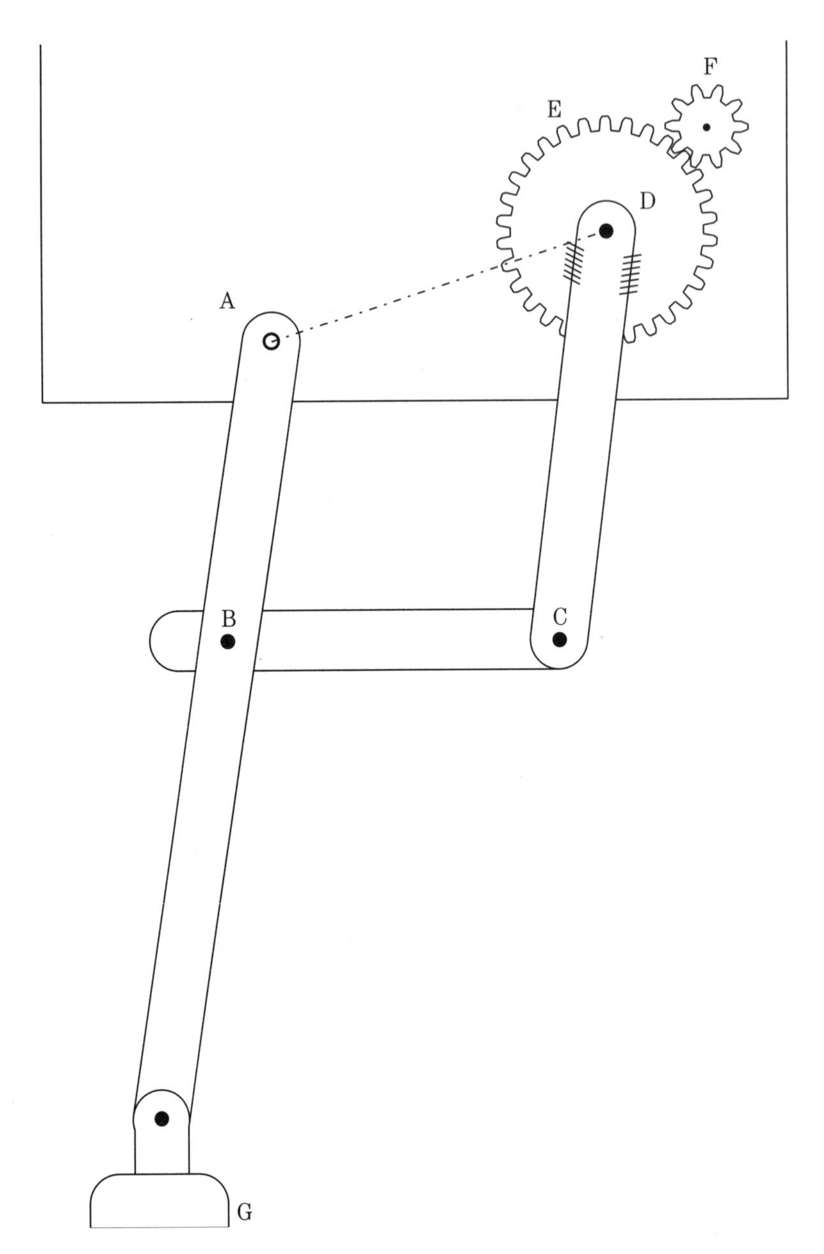

8-24 A four-bar linkage used as a leg mechanism for a robot.

Drag-link mechanism

When the shortest link is fixed as in Fig. 8-25(a), the linkage is called a *drag-link* mechanism and both **b** and **d** are cranks capable of rotating continuously through 360 degrees. If either **b** or **d** is rotated at a constant speed, the other will rotate in the same direction at a varying speed.

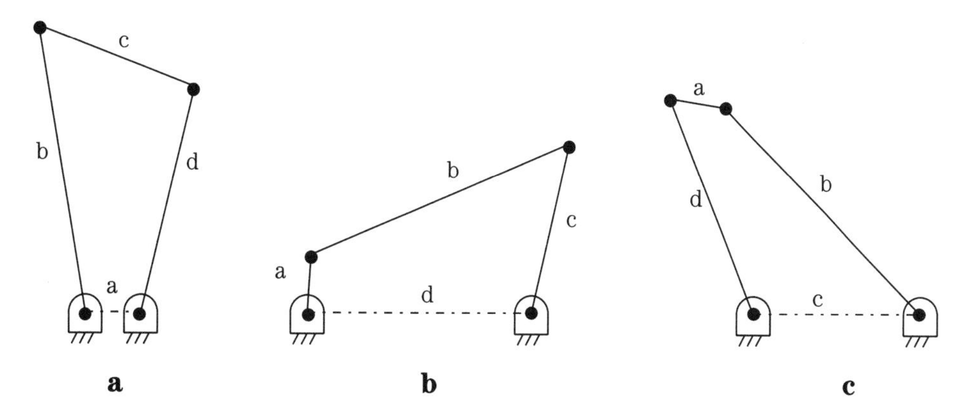

8-25 Three different inversions of a four-bar mechanism: (a) drag-link mechanism, (b) crank-rocker mechanism, (c) double-rocker mechanism.

Crank-rocker mechanism

When the shortest link is the driver as in Fig. 8-25(b), the linkage is called a *crank-rocker* mechanism. The driver can rotate continuously through 360 degrees (i.e., it is a crank), while the output link **c** can only oscillate (i.e., it is a rocker). The crank-rocker mechanism of Fig. 8-25(b) will have two *toggle* positions at which the rotational direction of link **c** reverses. As depicted in Fig. 8-26, these toggle positions occur when links **a** and **b** are colinear. One toggle position has **a** and **b** end-to-end as in Fig. 8-26(a), and the other has **a** and **b** overlapping as in Fig. 8-26(c).

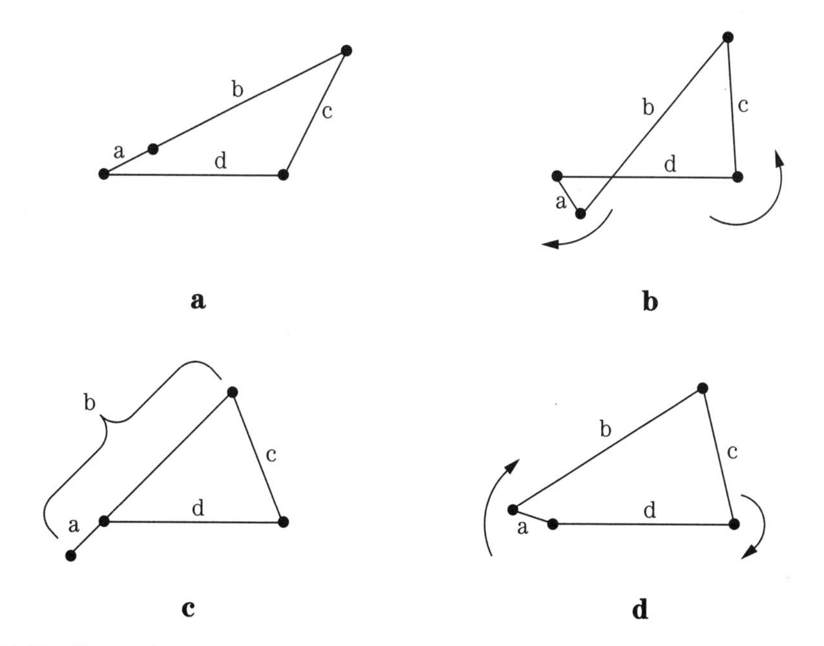

8-26 Toggling action of a crank-rocker mechanism: (a) toggle position with links **a** and **b** end-to-end, (b) rocker **c** rotating counterclockwise, (c) toggle position with links **a** and **b** overlapping, (d) rocker **c** rotating clockwise.

As link **a** rotates clockwise from the position of Fig. 8-26(a) to the position of Fig. 8-26(c), link **c** will rotate counterclockwise as shown in Fig. 8-26(b). As link **a** continues to rotate clockwise from the position of Fig. 8-26(c) to the position of Fig. 8-26(a), link **c** will reverse its direction, rotating clockwise as shown in Fig. 8-26(d) until the position of Fig. 8-26(a) is reached. Note that the two toggle positions are not 180 degrees apart in terms of link **a**'s position. To put this in concrete terms, let's assume that $a = 1$, $b = 5$, $c = 3$, and $d = 4$. For this specific case, link **a** will be at approximately 26.4° for the toggle position depicted in Fig 8-26(a) and at approximately 224.1° for the toggle position depicted in Fig. 8-26(c). Thus, when moving clockwise from the position of Fig. 8-26(c) to the position of Fig. 8-26(a), link **a** will rotate through 197.7°. If link **a** rotates clockwise at a constant speed, it will spend more time going from the position of Fig. 8-26(c) to the position of Fig. 8-26(a) than it spends going from Fig. 8-26(a) to Fig. 8-26(c). Thus, link **c** will spend more time going counterclockwise than going clockwise. This condition is sometimes called a *quick-return* feature of the linkage.

Double-rocker mechanism

When the shortest link is the coupler as in Fig. 8-25(c), the linkage is called a *double-rocker* mechanism, and both links **b** and **d** will be restricted to oscillation (*i.e.,* **b** and **d** are rockers). One extreme of the oscillation will be a dead point and the other extreme will be a toggle position. Which is which depends upon whether **b** or **d** is the driven link. When the Grashoff rule is not satisfied (*i.e.,* when $a + b > c + b$), the linkage will be a double-rocker mechanism, regardless of how the links are configured.

Standard analysis format

So far, we have used **a** to denote the length of the shortest link and **b** to denote the length of the longest link. These assignments helped prevent confusion as the roles of the different-length links were interchanged. In most other analyses of four-bar linkages, it is more convenient to denote the links as follows:

a = length of the fixed link

b = length of the input or driver link

c = length of the coupler

d = length of the output link

Furthermore, as shown in Fig. 8-27, let us denote the angles between the links as:

θ = clockwise angle from the fixed link to the input link

ϕ = clockwise angle from the fixed link to the output link

α = clockwise angle from the fixed link to the coupling link

The relationship between the input angle θ and output angle ϕ is expressed by the Freudenstein equation:

$$R_1 \cos\theta - R_2 \cos\theta + R_3 = \cos(\theta - \phi)$$

where

$$R_1 = \frac{a}{d}$$

$$R_2 = \frac{a}{b}$$

$$R_3 = \frac{a^2 + b^2 - c^2 + d^2}{2bd}$$

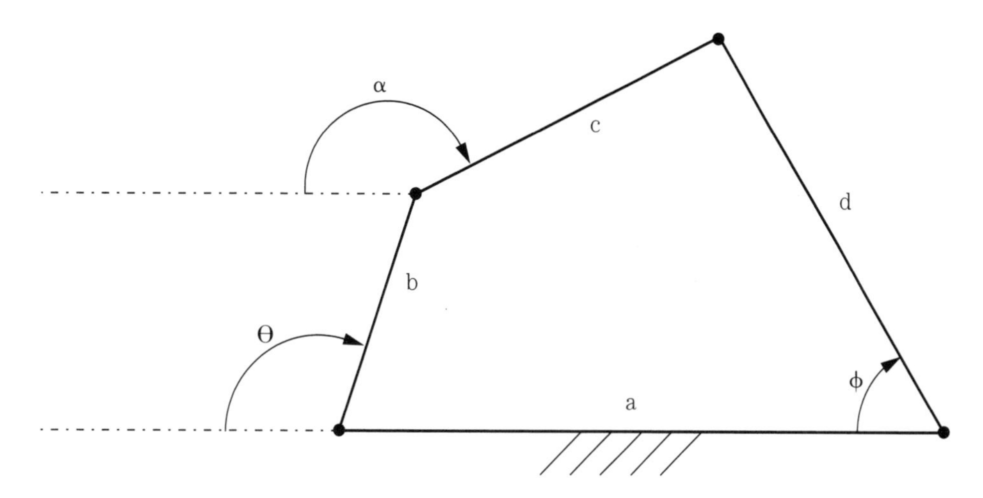

8-27 Standard format for analysis of four-bar linkage.

Linkage construction

Design of linkages can be a complicated affair. Mechanical engineers often must use computer-aided design programs and computer simulations to design a linkage and verify that it will operate as desired. For experimenters not having access to this sort of software, which can be quite expensive, there is an alternative approach consisting of manual design followed by experimental verification. Put simply, this experimental verification consists of testing a candidate design in some inexpensive and easily worked material before fabricating the final parts. For planar linkages, cardboard links that pivot on map pins stuck into a corkboard is my favorite way to confirm that a linkage will move the way I want it to before I go ahead and make the final parts in plastic or aluminum. For designs that involve pivots that move in slots, bolts with nuts and washers can be used as shown in Fig. 8-28.

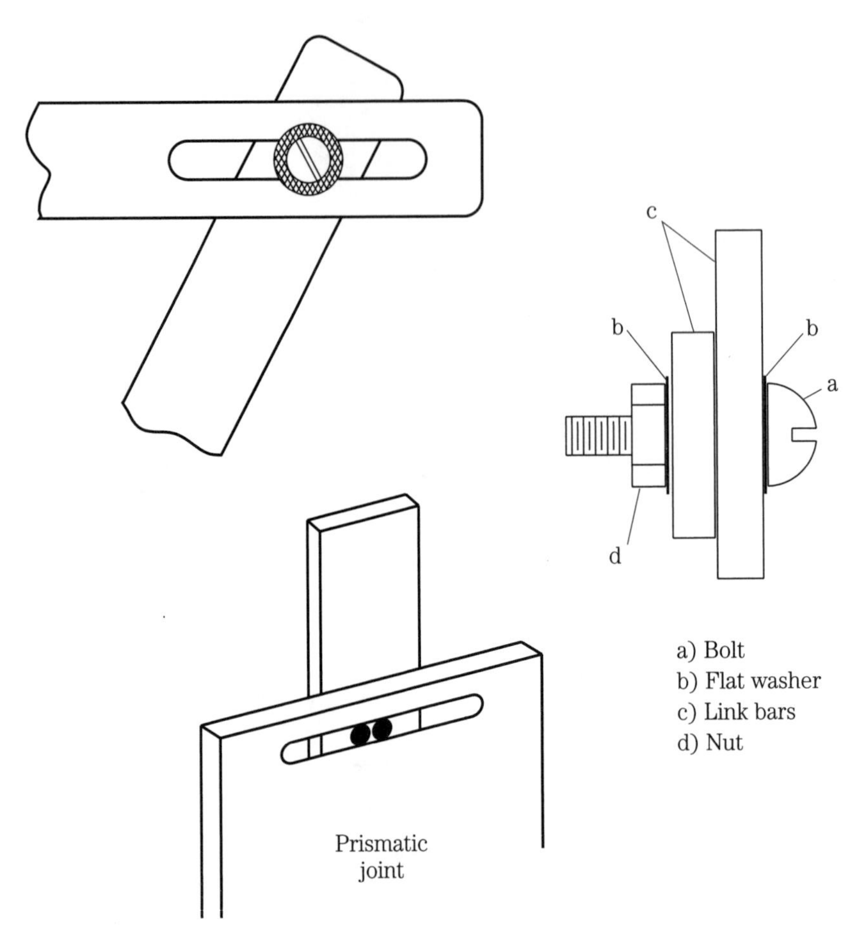

8-28 Methods for constructing joints in cardboard mock-ups of linkages.

9
CHAPTER
Pneumatic systems

Pneumatic systems use pressurized air to create motion and force in mechanical devices. These systems are usually less powerful than hydraulic systems of similar size, but they are much cleaner to service and operate. A small leak of hydraulic fluid on the living room carpeting is a mess; a small leak of compressed air is a "don't care." Pneumatic systems are best reserved for moderate force, moderate-precision applications. Electric motor drives are somewhat easier to control than pneumatics when small precision movements are called for. When large forces are needed, hydraulic systems are superior to pneumatic systems.

Air

At sea level, the normal air pressure is about 14.7 pounds per square inch (psi). There are two different ways that pressure in a pneumatic system can be reported. For the sake of example, consider a pressurized air tank in which the pressure is 18.0 psi. This can be reported as an *absolute* pressure of 18.0 psia, or as *gauge* pressure relative to ambient of 3.3 psig. The weather service reports the ambient atmospheric pressure as the *barometric* pressure. A mercury barometer is shown in Fig. 9-1.

The glass tube is completely filled with mercury, and then inverted with the open end below the surface of the mercury in the cistern as shown in the figure. The weight of the mercury will cause the column in the tube to drop below the closed end of the tube. The empty space this creates is **empty**; it does not even contain air. Air pressure on the surface of the cistern helps keep the column from dropping too far. The higher the pressure, the taller the column of mercury. The column will typically extend about 30 inches above the surface of the mercury in the cistern. In fair-weather high pressure systems, the column will be higher. During periods of stormy weather, the height of the column will be lower. A barometer reading of 29.92 inches corresponds to the "standard" atmospheric pressure of 14.7 psi. Usually scientists are more interested in absolute pressures, and engineers are more interested in gauge pressures.

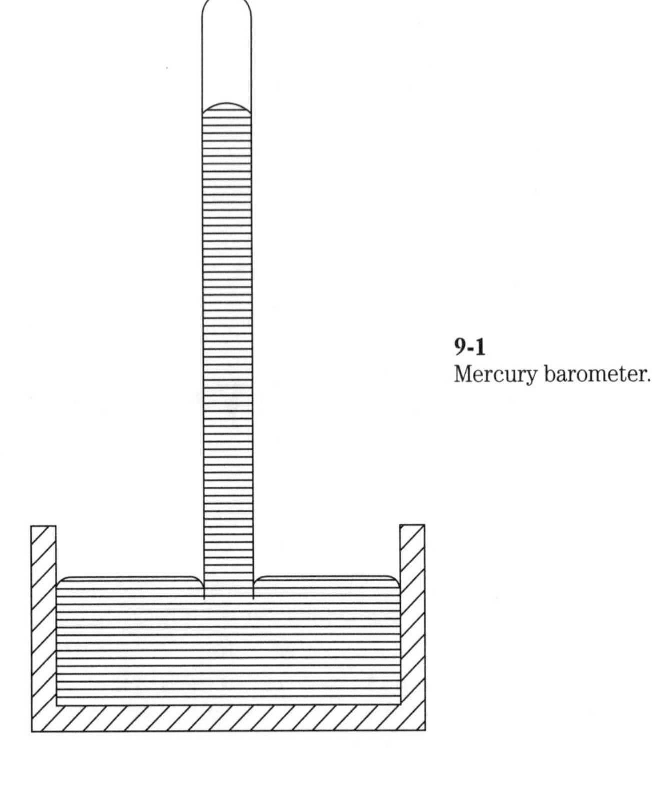

9-1
Mercury barometer.

Compression

Gases such as nitrogen, oxygen, and air are *compressible*. This means that when a gas is subjected to increased pressure, the volume occupied by the gas will decrease. Or conversely, if a fixed amount of gas is forced into a smaller volume, the pressure in the gas will increase. The temperature, pressure, and volume of an *ideal* gas are related in such a way that if one of these quantities is changed, the others will change such that:

$$\frac{P_1V_1}{T_1} = \frac{P_2V_2}{T_2} \tag{9-1}$$

where

P_1 = pressure before change
P_2 = pressure after change
V_1 = volume before change
V_2 = volume after change
T_1 = temperature (kelvin) before change
T_2 = temperature (kelvin) after change

Although air is not an ideal gas, Eq. 9-1 will be a reasonable approximation for pneumatic experimentation purposes.

Pneumatic components

The basic elements of a pneumatic system include a source of compressed air and some sort of device for creating force and/or motion from this compressed air.

Pistons

A cross section view of a cylinder and piston is shown in Fig. 9-2. If pressurized air is fed into the cylinder via the inlet connection, the force on the piston will be:

$$F = PA$$

where

P = pressure of the air

A = surface area of the piston exposed to the pressurized air

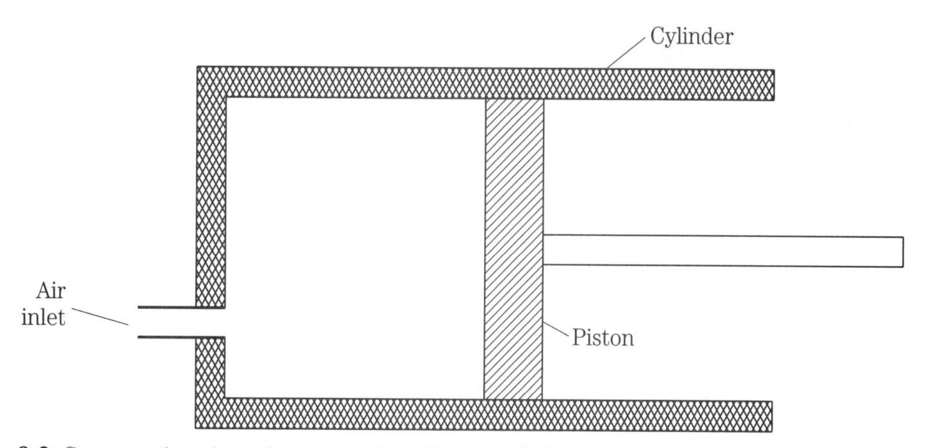

9-2 Cross section view of a pneumatic cylinder and piston.

Homemade piston

WARNING:

Safety should always come first, and **THIS IS ESPECIALLY IMPORTANT WHEN DEALING WITH HOMEMADE PNEUMATIC COMPONENTS.** An improperly constrained piston could be "launched" right out of its cylinder at speeds high enough to do serious damage to whatever or whomever it hits. **Never** subject homemade components to more than 15 or 20 psig. Inexpensive, easily obtained compressors can produce pressures much higher than this, so experimenters **must** be **extremely** careful to make sure that homemade pneumatic components are not accidentally subjected to higher pressures than intended. Another safety rule: **DON'T USE GLASS.**

Despite the dire warnings above, pneumatic systems **can** be used safely. Some LEGO Technics building sets intended for 10-year-olds come with the pneumatic components shown in Fig. 9-3. One feature of the LEGO system that guarantees safe

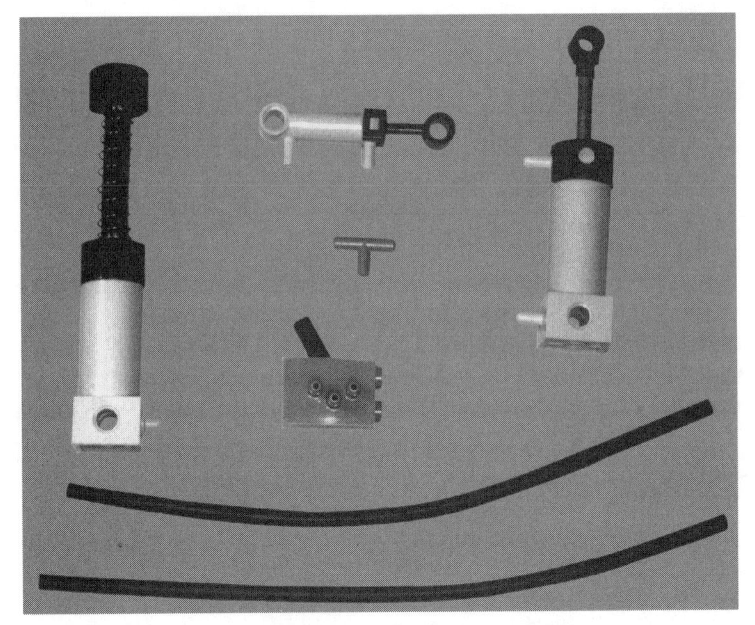

9-3 Pneumatic components from LEGO Technics building set.

operation is the way that the various components are connected. Small-diameter rubber tubing is simply forced over molded plastic inlet fittings on the various components.

If the pressure in the system gets too high, one of these connections will simply pop apart, long before any of the components even get close to their safe pressure limits. This would be a good idea for homebrew systems as well. Hose clamps or threaded fittings may be necessary in locations subject to a lot of motion, but have at least one "pop-able" connection in **each** link of the system. (I attempted to measure the pressure at which the tubing would pop off of the LEGO pump, but I was unable to get the pressure high enough. It seems that the pump has some sort of internal "blowby" feature that keeps the output pressure from getting too high.)

Reasonably air-tight cylinders and pistons can be fabricated in an amateur shop. In most cases, the cylinder itself will be some sort of rigid smooth-bore tubing or pipe. O-rings can be used to form a good seal between the piston and the cylinder. O-rings are circular rings of rubber having a circular cross section as shown in Fig. 9-4.

There are three measurements that we could make to characterize an O-ring. These measurements are the outside diameter (O.D.), inside diameter (I.D.), and cross-sectional diameter. Of course, any two of these measurements are enough to specify the ring, because:

$$d_o = d_i + 2d_{cs}$$

where

d_o = outside diameter
d_i = inside diameter
d_{cs} = cross-sectional diameter

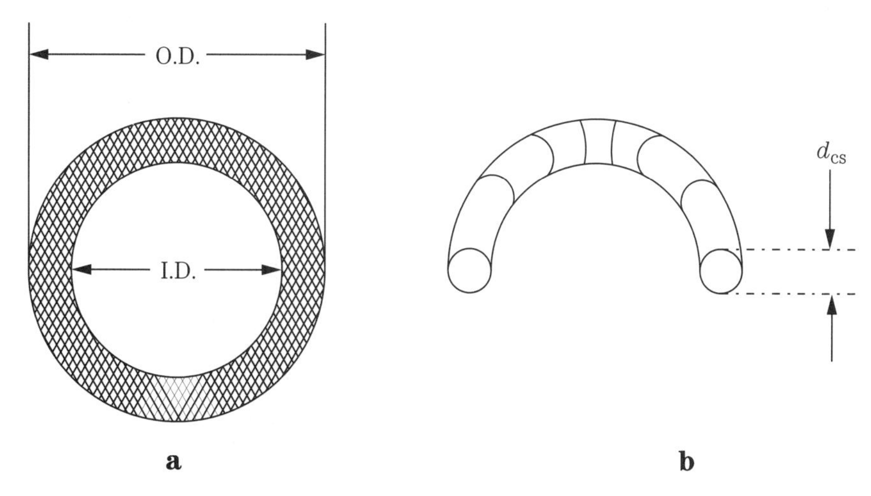

9-4 O-ring dimensioning (a) and cutaway (b) showing circular cross-section.

Figure 9-5 shows a cutaway view of how an O-ring can be used to form a seal between a piston and a cylinder. This is a method that is easily used in the home shop. The cylinder wall is a section of manufactured tubing, and the piston is a section of manufactured rod stock. The success of this method depends on:

- Selecting tubing, rod stock, and O-rings that have the proper size relationship
- The builder's ability to accurately cut a groove of proper dimension around the circumference of the piston material.

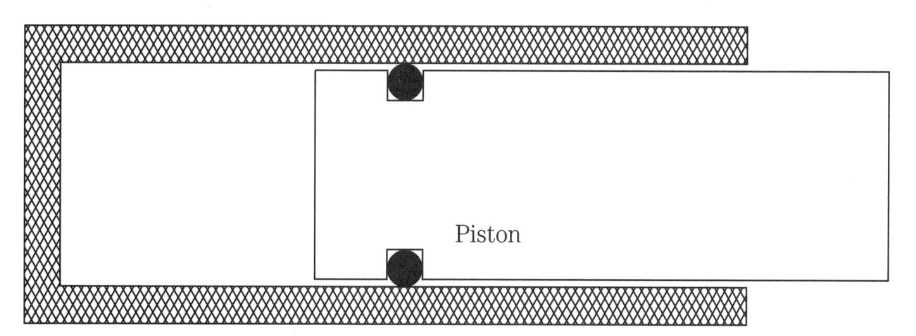

9-5 O-ring used to form a seal between a piston and a cylinder.

Just what are the proper size relationships? As shown in Fig. 9-6, the distance from the bottom of the piston groove to the cylinder wall must be smaller than the cross-sectional area of the O-rings, thus squeezing the O-ring into intimate contact with both the piston and cylinder. Standard O-rings are available with cross-sectional diameters ranging from 1⁄32 in. to 1⁄4 in. It has been determined that O-ring seals work best when the O-ring cross-sectional diameter is squeezed to about 90 percent of its original value. To make designers' lives easier, standard O-rings come with their actual cross-sectional diameters about 10 percent larger than their nominal cross-sectional diameters. Thus the space for the O-ring can be designed and machined to

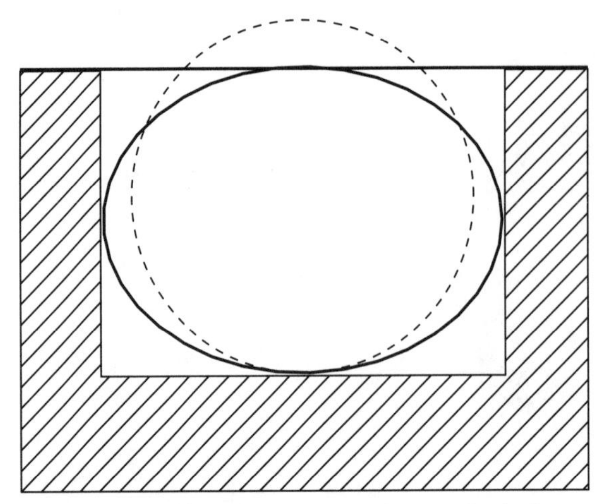

9-6 The O-ring must be squeezed to make intimate contact with both the piston and cylinder.

allow the exact amount of room that would be needed for the nominal size. Then, when an actual ring that is 10% oversize is installed in this space, it will automatically be squeezed by the proper amount.

Table 9-1.

	Nominal size			Actual dimensions	
Dash number	**I.D.**	**O.D.**	**Width**	**I.D.**	**Width**
–001	1/32	3/32	1/32	.029	.040
–001½	1/16	1/8	1/32	.070	.040
–002	3/64	9/64	3/64	.042	.050
–003	1/16	3/16	1/16	.056	.060
–004	5/64	13/64	1/16	.070	.070
–005	3/32	7/32	1/16	.101	.070
–006	1/8	1/4	1/16	.114	.070
–007	5/32	9/32	1/16	.145	.070
–008	3/16	5/16	1/16	.176	.070
–009	7/32	11/32	1/16	.208	.070
–010	1/4	3/8	1/16	.239	.070
–011	5/16	7/16	1/16	.301	.070
–012	3/8	1/2	1/16	.364	.070
–013	7/16	9/16	1/16	.426	.070
–014	1/2	5/8	1/16	.489	.070
–015	9/16	11/16	1/16	.551	.070
–016	5/8	3/4	1/16	.614	.070
–017	11/16	13/16	1/16	.676	.070
–018	3/4	7/8	1/16	.739	.070

Dash number	Nominal size I.D.	Nominal size O.D.	Nominal size Width	Actual dimensions I.D.	Actual dimensions Width
–019	13/16	15/16	1/16	.801	.070
–020	7/8	1	1/16	.864	.070
–021	15/16	1 1/16	1/16	.926	.070
–022	1	1 1/8	1/16	.989	.070
–023	1 1/16	1 3/16	1/16	1.051	.070
–024	1 1/8	1 1/4	1/16	1.114	.070
–025	1 3/16	1 5/16	1/16	1.176	.070
–026	1 1/4	1 3/8	1/16	1.239	.070
–027	1 5/16	1 7/16	1/16	1.301	.070
–028	1 3/8	1 1/2	1/16	1.364	.070
–029	1 1/2	1 5/8	1/16	1.489	.070
–030	1 5/8	1 3/4	1/16	1.614	.070
–031	1 3/4	1 7/8	1/16	1.739	.070
–032	1 7/8	2	1/16	1.864	.070
–033	2	2 1/8	1/16	1.989	.070
–034	2 1/8	2 1/4	1/16	2.114	.070
–035	2 1/4	2 3/8	1/16	2.239	.070
–036	2 3/8	2 1/2	1/16	2.364	.070
–037	2 1/2	2 5/8	1/16	2.489	.070
–038	2 5/8	2 3/4	1/16	2.614	.070
–039	2 3/4	2 7/8	1/16	2.739	.070
–040	2 7/8	3	1/16	2.864	.070
–041	3	3 1/8	1/16	2.989	.070
–042	3 1/4	3 3/8	1/16	3.239	.070
–043	3 1/2	3 5/8	1/16	3.489	.070
–044	3 3/4	3 7/8	1/16	3.739	.070
–045	4	4 1/8	1/16	3.989	.070
–046	4 1/4	4 3/8	1/16	4.239	.070
–047	4 1/2	4 5/8	1/16	4.489	.070
–048	4 3/4	4 7/8	1/16	4.739	.070
–049	5	5 1/8	1/16	4.989	.070
–050	5 1/4	5 3/8	1/16	5.239	.070
–102	1/16	1/4	3/32	.049	.103
–103	3/32	9/32	3/32	.081	.103
–104	1/8	5/16	3/32	.112	.103
–105	5/32	11/32	3/32	.143	.103
–106	3/16	3/8	3/32	.174	.103
–107	7/32	13/32	3/32	.206	.103
–108	1/4	7/16	3/32	.237	.103
–109	5/16	1/2	3/32	.299	.103

Table 9-1. Continued.

Dash number	Nominal size I.D.	Nominal size O.D.	Nominal size Width	Actual dimensions I.D.	Actual dimensions Width
–110	3/8	9/16	3/32	.362	.103
–111	7/16	5/8	3/32	.424	.103
–112	1/2	11/16	3/32	.487	.103
–113	9/16	3/4	3/32	.549	.103
–114	5/8	13/16	3/32	.612	.103
–115	11/16	7/8	3/32	.674	.103
–116	3/4	15/16	3/32	.737	.103
–117	13/16	1	3/32	.799	.103
–118	7/8	1 1/16	3/32	.862	.103
–119	15/16	1 1/8	3/32	.924	.103
–120	1	1 3/16	3/32	.987	.103
–121	1 1/16	1 1/4	3/32	1.049	.103
–122	1 1/8	1 5/16	3/32	1.112	.103
–123	1 3/16	1 3/8	3/32	1.174	.103
–124	1 1/4	1 7/16	3/32	1.237	.103
–125	1 5/16	1 1/2	3/32	1.299	.103
–126	1 3/8	1 9/16	3/32	1.362	.103
–127	1 7/16	1 5/8	3/32	1.424	.103
–128	1 1/2	1 11/16	3/32	1.487	.103
–129	1 9/16	1 3/4	3/32	1.549	.103
–130	1 5/8	1 3/16	3/32	1.612	.103
–131	1 11/16	1 7/8	3/32	1.674	.103
–132	1 3/4	1 15/16	3/32	1.737	.103
–133	1 13/16	2	3/32	1.799	.103
–134	1 7/8	2 1/16	3/32	1.862	.103
–135	1 15/16	2 1/8	3/32	1.925	.103
–136	2	2 3/16	3/32	1.987	.103
–137	2 1/16	2 1/4	3/32	2.050	.103
–138	2 1/8	2 5/16	3/32	2.112	.103
–139	2 3/16	2 3/8	3/32	2.175	.103
–140	2 1/4	2 7/16	3/32	2.237	.103
–141	2 5/16	2 1/2	3/32	2.300	.103
–142	2 3/8	2 9/16	3/32	2.362	.103
–143	2 7/16	2 5/8	3/32	2.425	.103
–144	2 1/2	2 11/16	3/32	2.487	.103
–145	2 9/16	2 3/4	3/32	2.550	.103
–146	2 5/8	2 13/16	3/32	2.612	.103
–147	2 11/16	2 7/8	3/32	2.675	.103
–148	2 3/4	2 15/16	3/32	2.737	.103
–149	2 13/16	3	3/32	2.800	.103

Dash number	Nominal size I.D.	Nominal size O.D.	Nominal size Width	Actual dimensions I.D.	Actual dimensions Width
–150	2⅞	3$\frac{1}{16}$	$\frac{3}{32}$	2.862	.103
–151	3	3$\frac{3}{16}$	$\frac{3}{32}$	2.987	.103
–152	3¼	3$\frac{7}{16}$	$\frac{3}{32}$	3.237	.103
–153	3½	3$\frac{11}{16}$	$\frac{3}{32}$	3.487	.103
–154	3¾	3$\frac{15}{16}$	$\frac{3}{32}$	3.737	.103
–155	4	4$\frac{3}{16}$	$\frac{3}{32}$	3.987	.103
–156	4¼	4$\frac{7}{16}$	$\frac{3}{32}$	4.237	.103
–157	4½	4$\frac{11}{16}$	$\frac{3}{32}$	4.487	.103
–158	4¾	4$\frac{15}{16}$	$\frac{3}{32}$	4.737	.103
–159	5	5$\frac{3}{16}$	$\frac{3}{32}$	4.987	.103
–160	5¼	5$\frac{7}{16}$	$\frac{3}{32}$	5.237	.103
–161	5½	5$\frac{11}{16}$	$\frac{3}{32}$	5.487	.103
–162	5¾	5$\frac{15}{16}$	$\frac{3}{32}$	5.737	.103
–163	6	6$\frac{3}{16}$	$\frac{3}{32}$	5.987	.103
–164	6¼	6$\frac{7}{16}$	$\frac{3}{32}$	6.237	.103
–165	6½	6$\frac{11}{16}$	$\frac{3}{32}$	6.487	.103
–166	6¾	6$\frac{15}{16}$	$\frac{3}{32}$	6.737	.103
–167	7	7$\frac{3}{16}$	$\frac{3}{32}$	6.987	.103
–168	7¼	7$\frac{7}{16}$	$\frac{3}{32}$	7.237	.103
–169	7½	7$\frac{11}{16}$	$\frac{3}{32}$	7.487	.103
–170	7¾	7$\frac{15}{16}$	$\frac{3}{32}$	7.737	.103
–171	8	8$\frac{3}{16}$	$\frac{3}{32}$	7.987	.103
–172	8¼	8$\frac{7}{16}$	$\frac{3}{32}$	8.237	.103
–173	8½	8$\frac{11}{16}$	$\frac{3}{32}$	8.487	.103
–174	8¾	8$\frac{15}{16}$	$\frac{3}{32}$	8.737	.103
–175	9	9$\frac{3}{16}$	$\frac{3}{32}$	8.987	.103
–176	9¼	9$\frac{7}{16}$	$\frac{3}{32}$	9.237	.103
–177	9½	9$\frac{11}{16}$	$\frac{3}{32}$	9.487	.103
–178	9¾	9$\frac{15}{16}$	$\frac{3}{32}$	9.737	.103
–201	$\frac{3}{16}$	$\frac{7}{16}$	⅛	.171	.139
–202	¼	½	⅛	.234	.139
–203	$\frac{5}{16}$	$\frac{9}{16}$	⅛	.296	.139
–204	⅜	⅝	⅛	.359	.139
–205	$\frac{7}{16}$	$\frac{11}{16}$	⅛	.421	.139
–206	½	¾	⅛	.484	.139
–207	$\frac{9}{16}$	$\frac{13}{16}$	⅛	.546	.139
–208	⅝	⅞	⅛	.609	.139
–209	$\frac{11}{16}$	$\frac{15}{16}$	⅛	.671	.139
–210	¾	1	⅛	.734	.139
–211	$\frac{13}{16}$	1$\frac{1}{16}$	⅛	.796	.139
–212	⅞	1⅛	⅛	.859	.139

Table 9-1. Continued.

Dash number	Nominal size I.D.	Nominal size O.D.	Nominal size Width	Actual dimensions I.D.	Actual dimensions Width
–213	15/16	1 3/16	1/8	.921	.139
–214	1	1 1/4	1/8	.984	.139
–215	1 1/16	1 5/16	1/8	1.046	.139
–216	1 1/8	1 3/8	1/8	1.109	.139
–217	1 3/16	1 7/16	1/8	1.171	.139
–218	1 1/4	1 1/2	1/8	1.234	.139
–219	1 5/16	1 9/16	1/8	1.296	.139
–220	1 3/8	1 5/8	1/8	1.359	.139
–221	1 7/16	1 11/16	1/8	1.421	.139
–222	1 1/2	1 3/4	1/8	1.484	.139
–223	1 5/8	1 7/8	1/8	1.609	.139
–224	1 3/4	2	1/8	1.734	.139
–225	1 7/8	2 1/8	1/8	1.859	.139
–226	2	2 1/4	1/8	1.984	.139
–227	2 1/8	2 3/8	1/8	2.109	.139
–228	2 1/4	2 1/2	1/8	2.234	.139
–229	2 3/8	2 5/8	1/8	2.359	.139
–230	2 1/2	2 3/4	1/8	2.484	.139
–231	2 5/8	2 7/8	1/8	2.609	.139
–232	2 3/4	3	1/8	2.734	.139
–233	2 7/8	3 1/8	1/8	2.859	.139
–234	3	3 1/4	1/8	2.984	.139
–235	3 1/8	3 3/8	1/8	3.109	.139
–236	3 1/4	3 1/2	1/8	3.234	.139
–237	3 3/8	3 5/8	1/8	3.359	.139
–238	3 1/2	3 3/4	1/8	3.484	.139
–239	3 5/8	3 7/8	1/8	3.609	.139
–240	3 3/4	4	1/8	3.734	.139
–241	3 7/8	4 1/8	1/8	3.859	.139
–242	4	4 1/4	1/8	4.984	.139
–243	4 1/8	4 3/8	1/8	4.109	.139
–244	4 1/4	4 1/2	1/8	4.234	.139
–245	4 3/8	4 5/8	1/8	4.359	.139
–246	4 1/2	4 3/4	1/8	4.484	.139
–247	4 5/8	4 7/8	1/8	4.609	.139
–248	4 3/4	5	1/8	4.734	.139
–249	4 7/8	5 1/8	1/8	4.859	.139

Dash number	Nominal size			Actual dimensions	
	I.D.	O.D.	Width	I.D.	Width
–250	5	5¼	⅛	4.984	.139
–251	5⅛	5⅜	⅛	5.109	.139
–252	5¼	5½	⅛	5.234	.139
–253	5⅜	5⅝	⅛	5.359	.139
–254	5½	5¾	⅛	5.484	.139
–255	5⅝	5⅞	⅛	5.609	.139
–256	5¾	6	⅛	5.734	.139
–257	5⅞	6⅛	⅛	5.859	.139
–258	6	6¼	⅛	5.984	.139
–259	6¼	6½	⅛	6.234	.139
–260	6½	6¾	⅛	6.484	.139
–261	6¾	7	⅛	6.734	.139
–262	7	7¼	⅛	6.984	.139
–263	7¼	7½	⅛	7.234	.139
–264	7½	7¾	⅛	7.484	.139
–265	7¾	8	⅛	7.734	.139
–266	8	8¼	⅛	7.984	.139
–267	8¼	8½	⅛	8.234	.139
–268	8½	8¾	⅛	8.484	.139
–269	8¾	9	⅛	8.734	.139
–270	9	9¼	⅛	8.984	.139
–271	9¼	9½	⅛	9.234	.139
–272	9½	9¾	⅛	9.484	.139
–273	9¾	10	⅛	9.734	.139
–274	10	10¼	⅛	9.984	.139
–275	10½	10¾	⅛	10.484	.139
–276	11	11¼	⅛	10.984	.139
–277	11½	11¾	⅛	11.484	.139
–278	12	12¼	⅛	11.984	.139
–279	13	13¼	⅛	12.984	.139
–280	14	14¼	⅛	13.984	.139
–281	15	15¼	⅛	14.984	.139
–282	16	16¼	⅛	15.955	.139
–283	17	17¼	⅛	16.955	.139
–284	18	18¼	⅛	17.955	.139
–309	7/16	13/16	3/16	.412	.210
–310	½	⅞	3/16	.412	.210
–311	9/16	15/16	3/16	.537	.210
–312	⅝	1	3/16	.600	.210

Table 9-1. Continued.

Dash number	Nominal size I.D.	Nominal size O.D.	Nominal size Width	Actual dimensions I.D.	Actual dimensions Width
–313	11/16	1 1/16	3/16	.662	.210
–314	3/4	1 1/8	3/16	.725	.210
–315	13/16	1 3/16	3/16	.787	.210
–316	7/8	1 1/4	3/16	.850	.210
–317	15/16	1 5/16	3/16	.912	.210
–318	1	1 3/8	3/16	.975	.210
–319	1 1/16	1 7/16	3/16	1.037	.210
–320	1 1/8	1 1/2	3/16	1.100	.210
–321	1 3/16	1 9/16	3/16	1.162	.210
–322	1 1/4	1 5/8	3/16	1.225	.210
–323	1 5/16	1 11/16	3/16	1.287	.210
–324	1 3/8	1	3/16	1.350	.210
–325	1 1/2	1 7/8	3/16	1.475	.210
–326	1 5/8	2	3/16	1.600	.210
–327	1 3/4	2 1/8	3/16	1.725	.210
–328	1 7/8	2 1/4	3/16	1.850	.210
–329	2	2 3/8	3/16	1.975	.210
–330	2 1/8	2 1/2	3/16	2.100	.210
–331	2 1/4	2 5/8	3/16	2.225	.210
–332	2 3/8	2 3/4	3/16	2.350	.210
–333	2 1/2	2 7/8	3/16	2.475	.210
–334	2 5/8	3	3/16	2.600	.210
–335	2 3/4	3 1/8	3/16	2.725	.210
–336	2 7/8	3 1/4	3/16	2.850	.210
–337	3	3 3/8	3/16	2.975	.210
–338	3 1/8	3 1/2	3/16	3.100	.210
–339	3 1/4	3 5/8	3/16	3.225	.210
–340	3 3/8	3 3/4	3/16	3.350	.210
–341	3 1/2	3 7/8	3/16	3.475	.210
–342	3 5/8	4	3/16	3.600	.210
–343	3 3/4	4 1/8	3/16	3.725	.210
–344	3 7/8	4 1/4	3/16	3.850	.210
–345	4	4 3/8	3/16	3.975	.210
–346	4 1/8	4 1/2	3/16	4.100	.210
–347	4 1/4	4 5/8	3/16	4.225	.210
–348	4 3/8	4 3/4	3/16	4.350	.210
–349	4 1/2	4 7/8	3/16	4.475	.210

Dash number	Nominal size I.D.	O.D.	Width	Actual dimensions I.D.	Width
–350	4⅝	5	3⁄16	4.600	.210
–351	4¾	5⅛	3⁄16	4.725	.210
–352	4⅞	5¼	3⁄16	4.850	.210
–353	5	5⅜	3⁄16	4.975	.210
–354	5⅛	5½	3⁄16	5.100	.210
–355	5¼	5⅝	3⁄16	5.225	.210
–356	5⅜	5¾	3⁄16	5.350	.210
–357	5½	5⅞	3⁄16	5.475	.210
–358	5⅝	6	3⁄16	5.600	.210
–359	5¾	6⅛	3⁄16	5.725	.210
–360	5⅞	6¼	3⁄16	5.850	.210
–361	6	6⅜	3⁄16	5.975	.210
–362	6¼	6⅝	3⁄16	6.225	.210
–363	6½	6⅞	3⁄16	6.475	.210
–364	6¾	7⅛	3⁄16	6.725	.210
–365	7	7⅜	3⁄16	6.975	.210
–366	7¼	7⅝	3⁄16	7.225	.210
–367	7½	7⅞	3⁄16	7.475	.210
–368	7¾	8⅛	3⁄16	7.725	.210
–369	8	8⅜	3⁄16	7.975	.210
–370	8¼	8⅝	3⁄16	8.225	.210
–371	8½	8⅞	3⁄16	8.475	.210
–372	8¾	9⅛	3⁄16	8.725	.210
–373	9	9⅜	3⁄16	8.975	.210
–374	9¼	9⅝	3⁄16	9.225	.210
–375	9½	9⅞	3⁄16	9.475	.210
–376	9¾	10⅛	3⁄16	9.725	.210
–377	10	10⅜	3⁄16	9.975	.210
–378	10½	10⅞	3⁄16	10.475	.210
–379	11	11⅜	3⁄16	10.975	.210
–380	11½	11⅞	3⁄16	11.475	.210
–381	12	12⅜	3⁄16	11.975	.210
–382	13	13⅜	3⁄16	12.975	.210
–383	14	14⅜	3⁄16	13.975	.210
–384	15	15⅜	3⁄16	14.975	.210
–385	16	16⅜	3⁄16	15.955	.210
–386	17	17⅜	3⁄16	16.955	.210
–387	18	18⅜	3⁄16	17.955	.210
–388	19	19⅜	3⁄16	18.955	.210
–389	20	20⅜	3⁄16	19.955	.210

Table 9-1. Continued.

Dash number	Nominal size I.D.	Nominal size O.D.	Nominal size Width	Actual dimensions I.D.	Actual dimensions Width
–390	21	21⅜	3/16	20.955	.210
–391	22	22⅜	3/16	21.955	.210
–392	23	23⅜	3/16	22.940	.210
–393	24	24⅜	3/16	23.940	.210
–394	25	25⅜	3/16	24.940	.210
–395	26	26⅜	3/16	25.940	.210
–425	4½	5	¼	4.475	.275
–426	4⅝	5⅛	¼	4.600	.275
–427	4¾	5¼	¼	4.725	.275
–428	4⅞	5⅜	¼	4.850	.275
–429	5	5½	¼	4.975	.275
–430	5⅛	5⅝	¼	5.100	.275
–431	5¼	5¾	¼	5.225	.275
–432	5⅜	5⅞	¼	5.350	.275
–433	5½	6	¼	5.475	.275
–434	5⅝	6⅛	¼	5.600	.275
–435	5¾	6¼	¼	5.725	.275
–436	5⅞	6⅜	¼	5.850	.275
–437	6	6½	¼	5.975	.275
–438	6¼	6¾	¼	6.225	.275
–439	6½	7	¼	6.475	.275
–440	6¾	7¼	¼	6.725	.275
–441	7	7½	¼	6.975	.275
–442	7¼	7¾	¼	7.225	.275
–443	7½	8	¼	7.475	.275
–444	7¾	8¼	¼	7.725	.275
–445	8	8½	¼	7.975	.275
–446	8½	9	¼	8.475	.275
–447	9	9½	¼	8.975	.275
–448	9½	10	¼	9.475	.275
–449	10	10½	¼	9.975	.275
–450	10½	11	¼	10.475	.275
–451	11	11½	¼	10.975	.275
–452	11½	12	¼	11.475	.275
–453	12	12½	¼	11.975	.275
–454	12½	13	¼	12.475	.275
–455	13	13½	¼	12.975	.275
–456	13½	14	¼	13.475	.275

Dash number	Nominal size I.D.	O.D.	Width	Actual dimensions I.D.	Width
–457	14	14½	¼	13.975	.275
–458	14½	15	¼	14.475	.275
–459	15	15½	¼	14.975	.275
–460	15½	16	¼	15.475	.275
–461	16	16½	¼	15.955	.275
–462	16½	17	¼	16.455	.275
–463	17	17½	¼	16.955	.275
–464	17½	18	¼	17.455	.275
–465	18	18½	¼	17.955	.275
–466	18½	19	¼	18.455	.275
–467	19	19½	¼	18.955	.275
–468	19½	20	¼	19.455	.275
–469	20	20½	¼	19.955	.275
–470	21	21½	¼	20.955	.275
–471	22	22½	¼	21.955	.275
–472	23	23½	¼	22.940	.275
–473	24	24½	¼	23.940	.275
–474	25	25½	¼	24.940	.275
–475	26	26½	¼	25.940	.275

Nominal and actual dimensions for the cross-sectional diameters of standard O-rings are listed in Table 9-1. In industry, O-rings are usually specified using the uniform dash number system in which the numbers from –001 to –475 are used to identify the specific dimensions of O-rings. A table of these numbers and the corresponding dimensions is provided in appendix A. This table tells us what sizes can be obtained. To find out what sizes can be easily obtained, take a trip to your local plumbing supplier, buy a package of assorted O-rings, and make some measurements. As far as the tubing and rod stock are concerned, the rod stock should be as fat as possible while still being able to slide easily within the tubing. If the rod stock is too thin, the O-ring can get stuck in the gap between the rod and the inner tubing wall as shown in Fig. 9-7, thus causing the piston to bind.

The **actual** I.D. of the tubing must equal the **nominal** O.D. of the O-ring. The groove in the piston is cut to a depth such that the diameter measured groove bottom to groove bottom equals the **nominal** I.D. of the O-ring. (See Fig. 9-8.) The depth of the groove is given by:

$$h = \frac{(D - d)}{2}$$

where

D = actual diameter of the rod stock

d = nominal I.D. of the O-ring

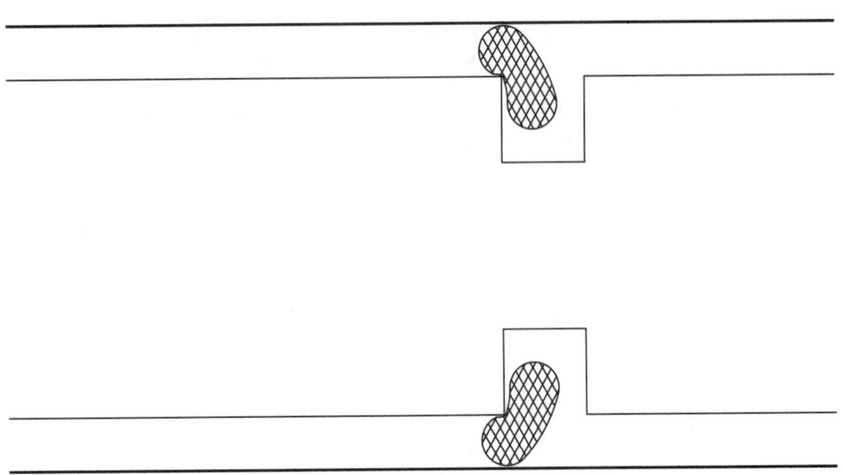

9-7 If the rod stock used for the piston is too small for the cylinder I.D., the O-ring can get stuck in the gap between the piston and cylinder.

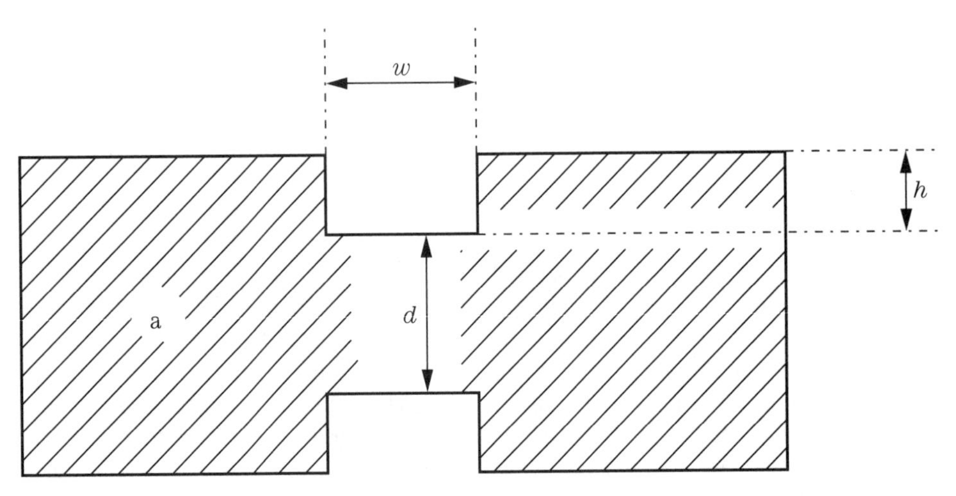

9-8 Dimensions of rod stock and groove that are used to select the correct size O-ring.

The width of the groove is usually between 1.5 and 2 times the cross-sectional diameter of the O-ring.

Metal pipe usually will not be suitable material for the cylinder. Iron pipe is too rough inside and copper pipe is too easily deformed into an out-of-round condition. For some miniature applications, the larger sizes of brass model airplane tubing might be usable. Transparent plastic has the advantage of letting you observe the internal operation of the piston. Of course, for high pressure applications thin-wall plastic would be a poor choice. Thick-wall PVC pipe can be used for larger cylinders. The piston itself can be made out of any material into which the required groove can be easily cut.

Door closer piston

Many pneumatically damped door closers such as the one shown in Fig. 9-9 can be modified for use as a pneumatic cylinder. These closers are equipped with an adjustment screw, which is usually located in the end of the cylinder as shown in Fig. 9-10. To test a cylinder to see if it can be used as a pneumatic cylinder, pull the rod halfway out and tighten the adjustment screw as far as it will go while holding the rod in the extended position. If the rod remains in the extended position, even when you apply moderate force to push it back in, the closer is suitable for use as a cylinder. If the rod slowly returns to its original position, the internal seals are leaking, and you should look for a better closer.

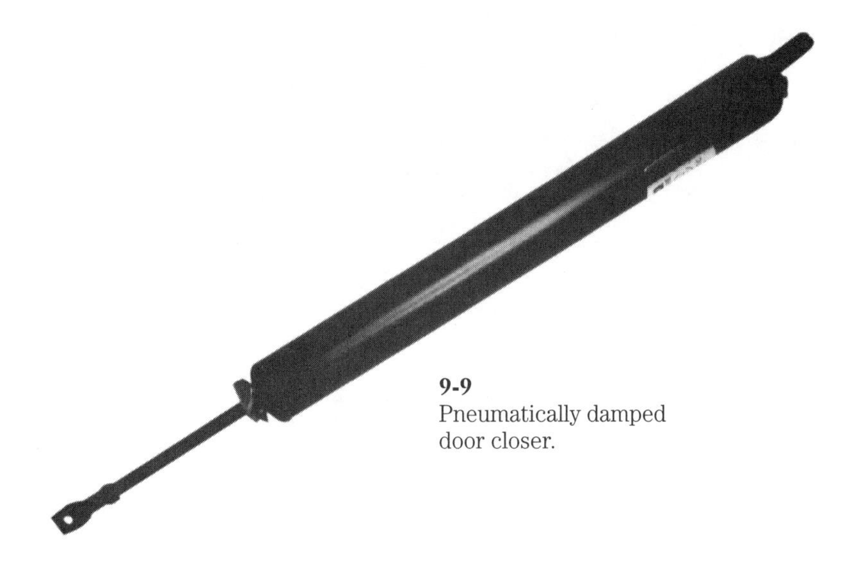

9-9
Pneumatically damped door closer.

The required modification is simple. Completely remove the adjustment screw and then enlarge and thread the vacated hole to accommodate a fitting that will allow an air hose to be connected to the cylinder. Two suitable connectors are shown in Fig. 9-11. Each fitting has a hose nipple on one end and male threads on the other end. The thinner fitting has 5/16-32 NEF (National Extra Fine) threads that match the threads on a standard tire stem. National Coarse (NC) and National Fine (NF) taps are readily available, but NEF taps are virtually **impossible** to find in a retail source. The fatter fitting has 1/8 male pipe threads. Taps for this thread are much easier to obtain.

Compressors

A common bicycle pump is an example of a reciprocating compressor that could actually prove usable in some applications. However, it is more likely that robot builders and other experimenters would prefer an electrically operated source of compressed air.

Compressors for operating paint sprayers and impact wrenches are sold by Sears and larger automotive supply stores. There are a number of specifications provided for such compressors, but the two of most interest (beside cost that is) are the air pressure and the *air rate*. Pneumatic tools require a certain air pressure in order

9-10 Closeup of door closer showing position of adjustment screw.

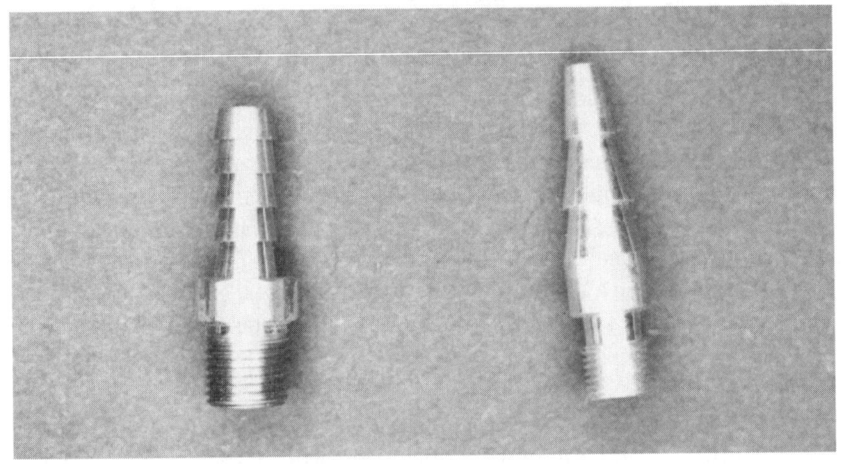

9-11 Two air fittings that can be used in modification to a pneumatic door closer.

to operate, and during normal operation, some of the pressurized air is vented to the atmosphere. This vented air must be replaced on the high pressure side of the tool. The amount of such "replacement" air that a compressor is capable of providing per minute is the air rate. Air rate is usually specified in *standard cubic feet per minute* (scfm). Typical compressor/tank combinations available to the home mechanic range from low-end ½ horsepower units that deliver 5 scfm at 15 psi to 20 hp units that deliver 100 scfm at 120 psi.

Tire inflators

There are also a wide variety of battery-powered compressors designed to be plugged into a car's cigarette lighter and used to inflate tires. These units are capable of working up to a pressure of 40 or 50 psig, but the rate of air flow is very low. It can take as long as five minutes to pump up a soft tire from 28 psi to 35 psi. Such units are in no way suitable for operating pneumatic tools, but in conjunction with the right reservoir they could be ideal for some small low-pressure pneumatic actuators in a robotic application. Being small and battery-powered, such units could form the basis for a pneumatic system that can be carried completely onboard a mobile platform.

Pressure reservoirs

When an air compressor is used to operate a pneumatic tool, there will be idle periods when the tool is not venting any air and the flow from the compressor to the tool is virtually zero. While the tool is being used, however, the demand for the compressor to supply air will be high. Using a *pressure reservoir* is the usual way to smooth out the highs and lows of demand and ensure that the tool will get a continuous supply of air when needed. This is the "tank" part of the compressor outfits sold for home use. This is also the dangerous part. Never use a reservoir tank that shows any signs of rust or damage.

Pressure gauges

The best source of ready-to-use pressure gauges is the automotive supply store. One popular design is a combination vacuum gauge and pressure gauge. The 4" circular combination gauge that I have is perfect for low-pressure homebrew systems. On the pressure side the needle travels through 130 degrees for a range of 0 to 10 psi. On the vacuum side, the needle travels through 210 degrees for a range of 0 to 30 inches of mercury.

Homemade pressure gauges

Figure 9-12 shows a device for measuring pressure differences. The tall U-shaped tube can be filled with water, hydraulic oil, or mercury. "Official" manometers used in professional laboratories are usually made of a continuous length of glass tubing that has been heated and bent into the required U-shape. However, a perfectly serviceable manometer can be made out of flexible clear plastic tubing held in the required shape with insulated staples of the sort sold for hanging Christmas lights. A more elegant solution might be to use a router to cut a U-shaped groove into a piece of 1×8 lumber and glue the flexible tubing in this groove. For a given pressure difference, heavier fluids (like mercury) will show a smaller difference in heights between the two legs of the U. Lighter fluids (like water) will show a larger difference. When very small differences in pressure need to be measured, a *slant-tube* manometer like the one in Fig. 9-13 can be used. Small changes in the vertical height of the liquid column will be easier to see, because vertical movements of the liquid will be accompanied by horizontal movements within the slanted tube. If the pressure difference between the two legs of any manometer is too large, the fluid can be blown completely out of the tube. Because of its greater sensitivity, a slant-tube manometer is especially prone to this sort of mishap.

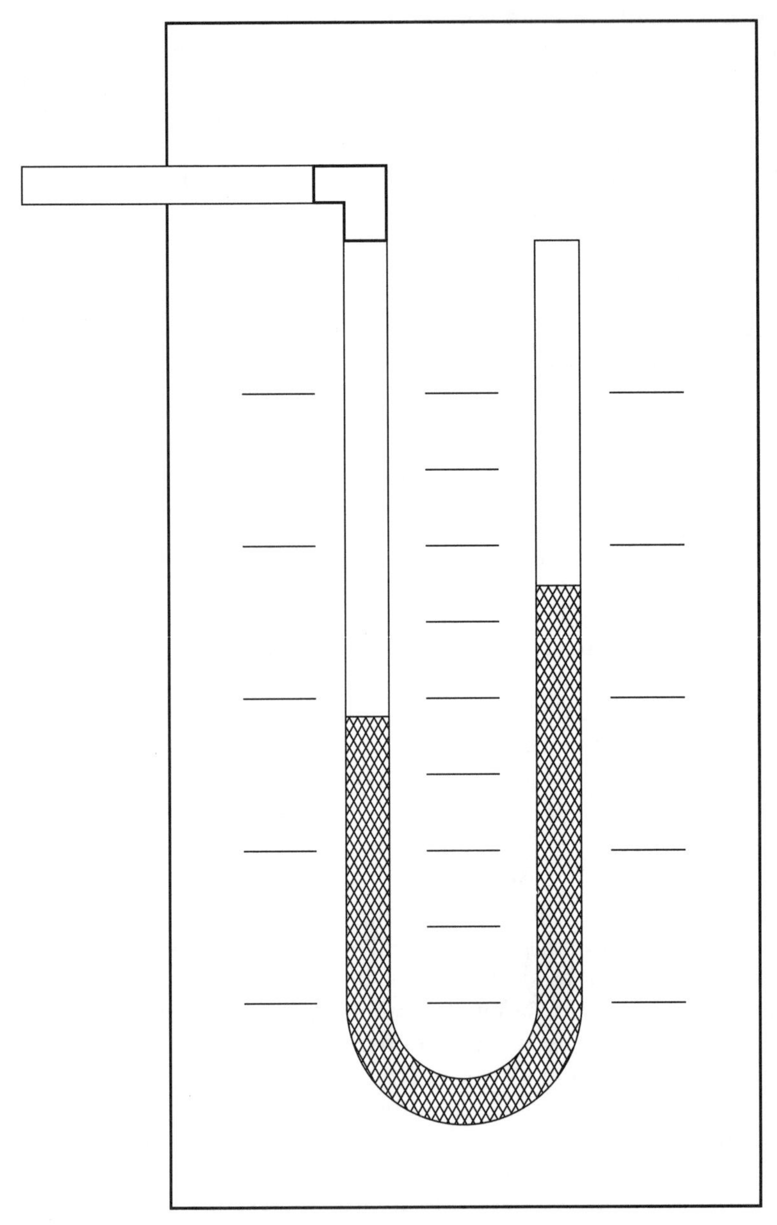

9-12 Manometer.

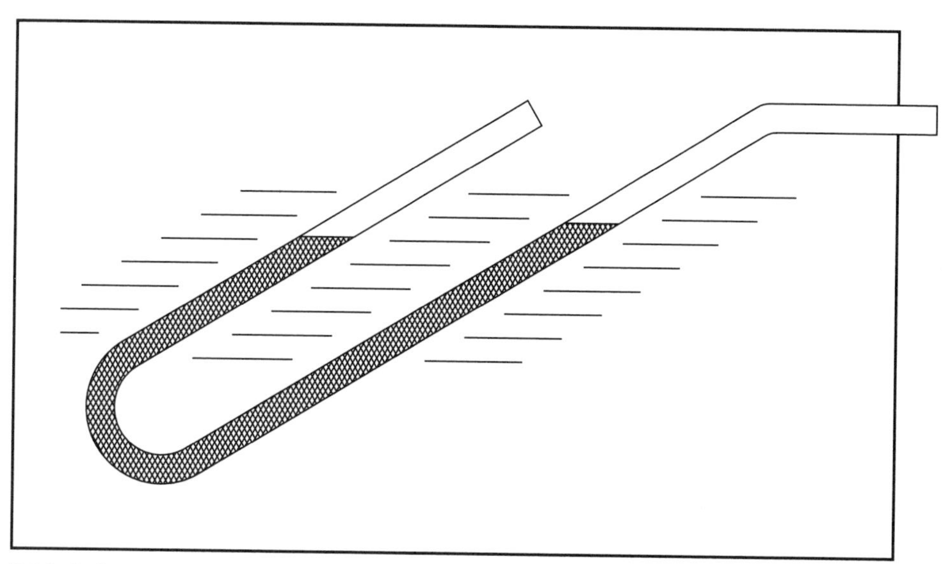

9-13 A slant-tube manometer.

10

CHAPTER

Vacuum systems

Vacuum systems have some of the same advantages that pneumatic systems have plus some of their own. Both vacuum systems and pneumatic systems have the advantage that leakage doesn't contaminate the environment. I'm not talking about saving the rain forest or closing the hole in the ozone layer; I am talking about loose hydraulic fluid in the environment of your living room. A vacuum can be released by venting it to the atmosphere, so there is no need to maintain a closed system as there is for hydraulics.

The pressure difference that can be developed in a vacuum system is hard-limited to about 14.7 psi, which is the difference between normal ambient atmospheric pressure outside of a vacuum component and zero pressure inside the vacuum component. No vacuum pump in the world can get below zero pressure, so any vessel that can withstand a 14.7 psi inward pressure will be safe no matter how good a vacuum pump is or how long it is allowed to operate. In pneumatic systems, pressure regulators are needed to ensure that compressors do not create more pressure than a vessel can withstand. On the down side, this limited pressure difference of 14.7 psi makes it more difficult to develop high forces in a compact volume with vacuum components than it is with either pneumatic components or hydraulic components. Pneumatic or hydraulic systems are inherently better than vacuum systems for applications such as robot legs, in which appendages must be forcefully extended and remain extended under compressive loads. Vacuum actuators are more naturally suited to resist tensile loading rather than compressive loading.

Vacuum pumps

Vacuum pumps can be either electrically operated or manually operated. For robotic applications, electrical vacuum pumps are the most convenient. Electrical pumps for use in portable robots should operate on battery power, but pumps requiring 110-Vac household power are okay for stationary projects, such as a pedestal-mounted arm and gripper system. Manually operated pumps are useful while experimenting and testing different vacuum-system configurations. Small hand-operated vacuum pumps like the one shown in Fig. 10-1 are sold in automo-

10-1
Hand-operated vacuum pump sold for testing automotive vacuum systems.

tive stores for use in testing various components in a car's vacuum system. Figure 10-2 shows a cross-sectional view of a typical bicycle pump. Many bicycle pumps can be converted into vacuum pumps by reversing the cup and valve as shown in Fig. 10-3.

Electric vacuum pumps are a bit more difficult. "Official" vacuum pumps are sold by scientific supply companies, but these pumps are very expensive; prices range from $250 to $1000. Because hand-operated bicycle pumps can so easily be converted into hand-operated vacuum pumps, I had always assumed (hoped?) that it would be easy to obtain a battery-operated vacuum pump by converting the inexpensive tire inflators that plug into a car's cigarette lighter. I took apart several different models of tire inflators and found their innards very similar, and not easily modifiable. On the other hand, it appears that the hand-operated unit shown in Fig. 10-1 can be easily modified for motorized operation. The body unscrews into three sections as shown in Fig. 10-4.

The return spring located between the piston and the handle is way too strong for motorized operation. In fact, it is probably not necessary to have any return spring at all. Change the spring or remove it entirely before reassembling the unit. Drill out the rivet that attaches the moveable handle to the fixed handle, and cut off the fixed handle just above the top of the rubber grip. Drill a mounting hole in the portion of handle still attached to the cylinder and mount the cylinder to a support as shown in Fig. 10-5.

The piston can be driven by a reciprocating push rod that is attached to a belt-driven wheel as shown in the figure. (The axle supports for the large wheel are not shown.) The total travel of the piston is slightly less than one inch, so the push rod pivot on the large wheel should be at a radius of slightly less than one-half inch. The large wheel should be positioned so that when the pivot is closest to the pump, the piston is full-forward in the cylinder. Likewise, when the pivot is farthest from the pump, the piston should be full-aft in the cylinder. It would be a good idea to use slots rather than holes for mounting the large wheel axle supports to the base. If the positioning is not quite right, the mounting screws can be loosened and the positioning adjusted before the mounting screws are retightened.

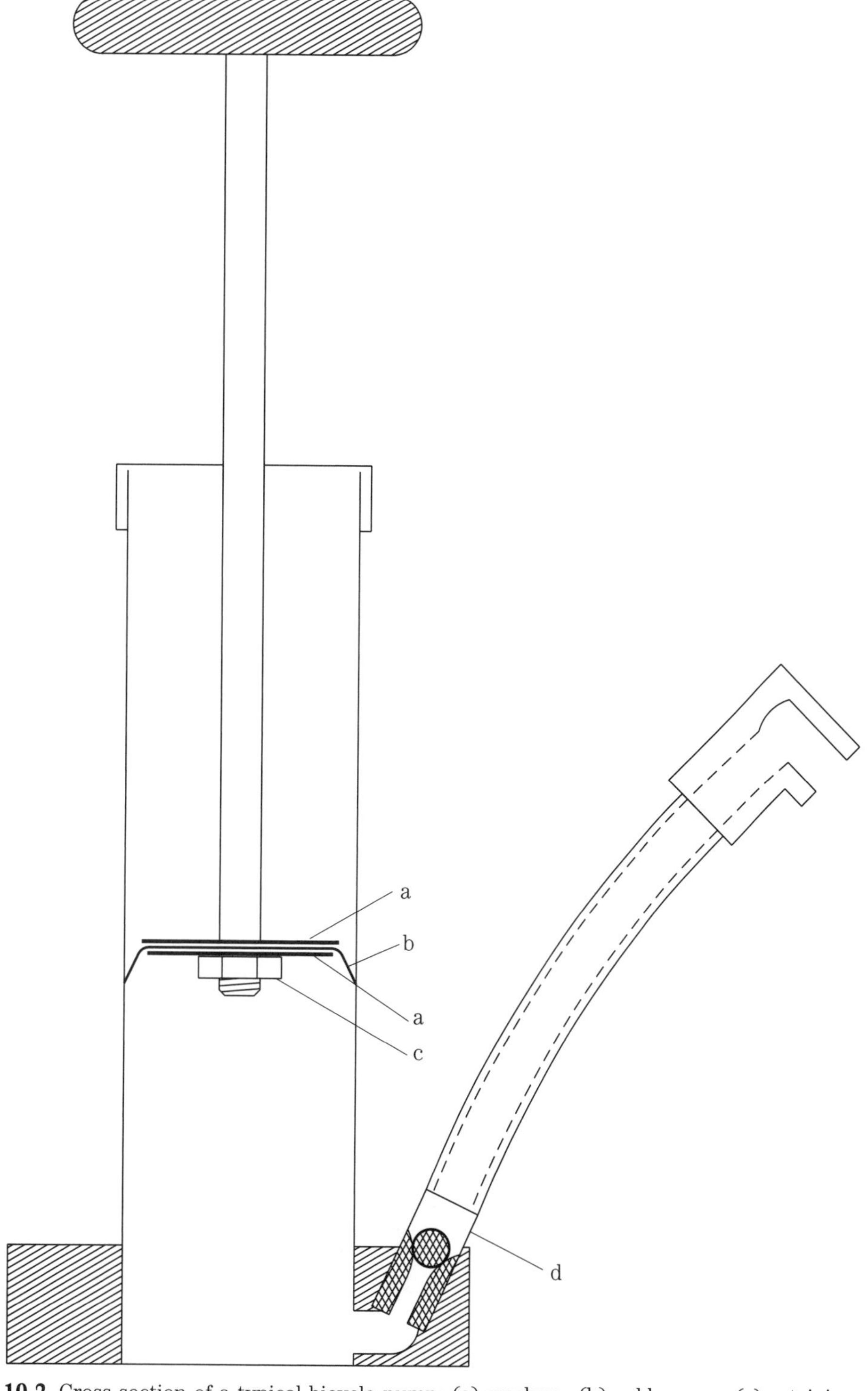

10-2 Cross-section of a typical bicycle pump: (a) washers, (b) rubber cup, (c) retaining nut, (d) check valve.

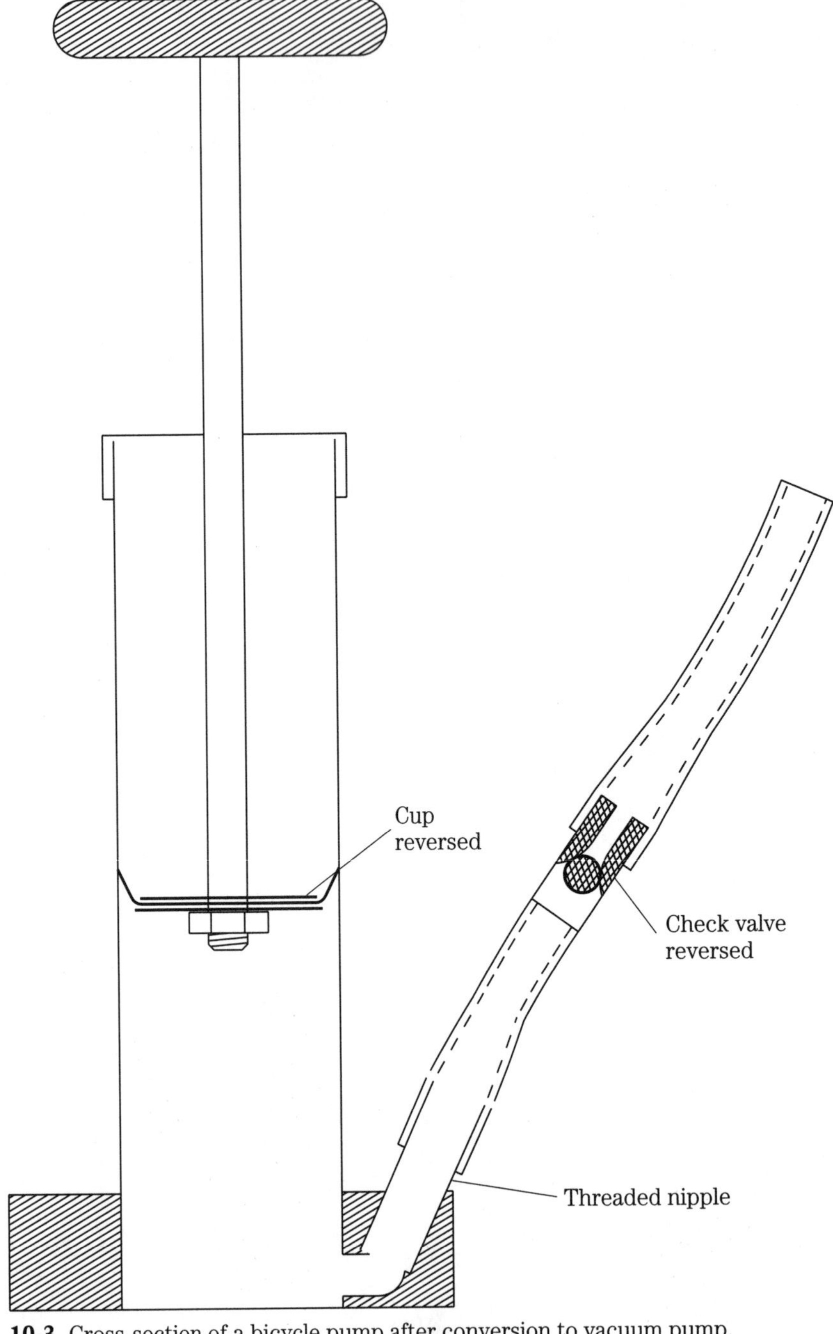

10-3 Cross-section of a bicycle pump after conversion to vacuum pump.

10-4 Hand-operated vacuum pump disassembled into three sections.

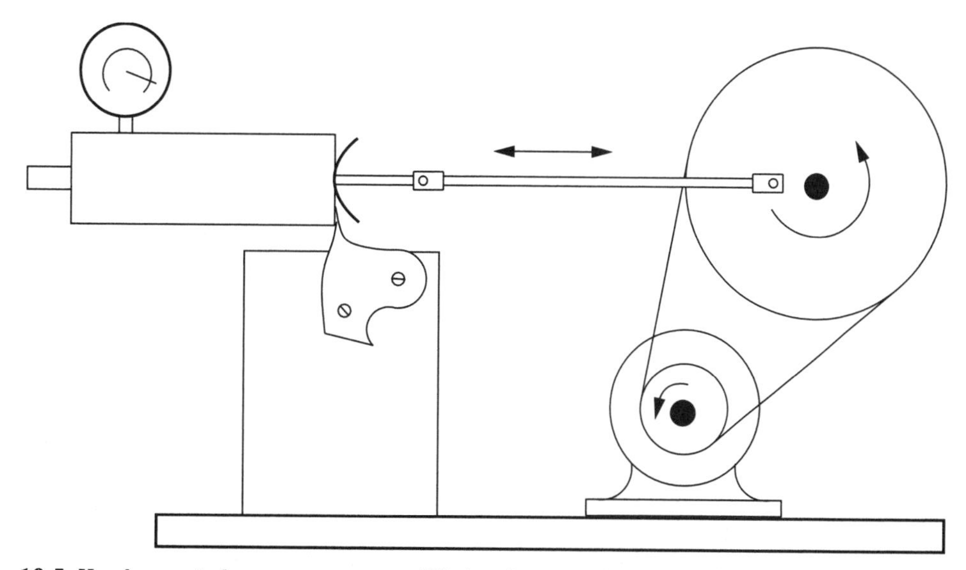

10-5 Hand-operated vacuum pump modified and mounted for motorized operation.

Vacuum actuators

It is possible to construct vacuum-operated pistons that are similar to pneumatically operated pistons. In our discussions of pneumatic pistons in chapter 9, we calculated the *net* force available from the piston as the product of the piston's face area and the **gauge** pressure of the compressed air in the cylinder.

$$F_{\text{net}} = A \times P_{\text{gauge}} \tag{10-1}$$

An alternative way to calculate the force would have been to compute the net force as the difference between the **total** force on the inside face of the piston and the force exerted by ambient atmospheric pressure on the outside of the piston.

$$F_{net} = F_{inside} - F_{outside} \tag{10-2}$$

The total force on the inside of the piston is the product of the piston's face area and the **absolute** pressure of the compressed air in the cylinder.

$$F_{inside} = A \times P_{absolute} \tag{10-3}$$

The force exerted by the atmosphere on the outside of the piston is the product of the piston's face area and the **ambient** atmospheric pressure.

$$F_{outside} = A \times P_{ambient} \tag{10-4}$$

Combining Eqs. 10-2, 10-3, 10-4, and making use of the fact that $P_{gauge} = P_{absolute} - P_{ambient}$, we obtain

$$\begin{aligned} F_{net} &= (A \times P_{absolute}) - (A \times P_{ambient}) \\ &= A \times (P_{absolute} - P_{ambient}) \\ &= A \times P_{gauge} \end{aligned}$$

This result agrees with Eq. 10-1. We can increase the net available force by increasing either the piston area A or the pressure difference $(P_{absolute} - P_{ambient})$.

Equation 10-2 is written for pneumatic pistons in which we assume that the inside pressure (and hence the inside force pushing the piston out) will always equal or exceed the outside pressure. For vacuum pistons, the opposite will be true: the outside force pushing in will always equal or exceed the inside force pushing out. Therefore if we want to avoid having to set up an algebraic sign convention for dealing with negative forces, we can rewrite Eq. 10-2 for vacuum pistons as:

$$\begin{aligned} F_{net} &= F_{outside} - F_{inside} \\ &= A \times (P_{ambient} - P_{absolute}) \end{aligned} \tag{10-5}$$

One way to increase the net force is to increase the difference between $P_{ambient}$ and $P_{absolute}$. Since $P_{ambient}$ will always be limited to about 14.7 psia, and $P_{absolute}$ inside the cylinder can never be less than zero, the maximum difference will be limited to about 14.7 psi. Once we are at or near this maximum pressure difference, the only other way to increase the available force is to increase the piston area. High forces will require large piston areas and consequently larger volumes for the complete piston-cylinder assembly. This is why the introduction notes that it is more difficult to develop high forces in a compact volume with vacuum components than it is with either pneumatic components or hydraulic components.

Hybrid pistons

In applications that call for large extension forces and relatively small retraction forces, a pneumatic/vacuum hybrid piston can be substituted for a double-acting pneumatic piston or a single-acting pneumatic piston with a spring return. The type of piston that uses an O-ring seal (as discussed in chapter 9) would be ideal for use as a hybrid piston, because the O-ring seal exhibits only slight directionality. The

O-ring will tend to move to the "inboard" (*i.e.*, towards the cylinder interior) side of the piston groove when the piston is moving out of the cylinder. When the piston is retracted into the cylinder, the O-ring will tend to move to the "outboard" side of the piston groove. This will happen whenever the piston is retracted regardless of how this retraction is accomplished, and it should not present any special obstacles to using a vacuum in the cylinder to retract the piston. A schematic diagram of a hybrid piston system is shown in Fig. 10-6.

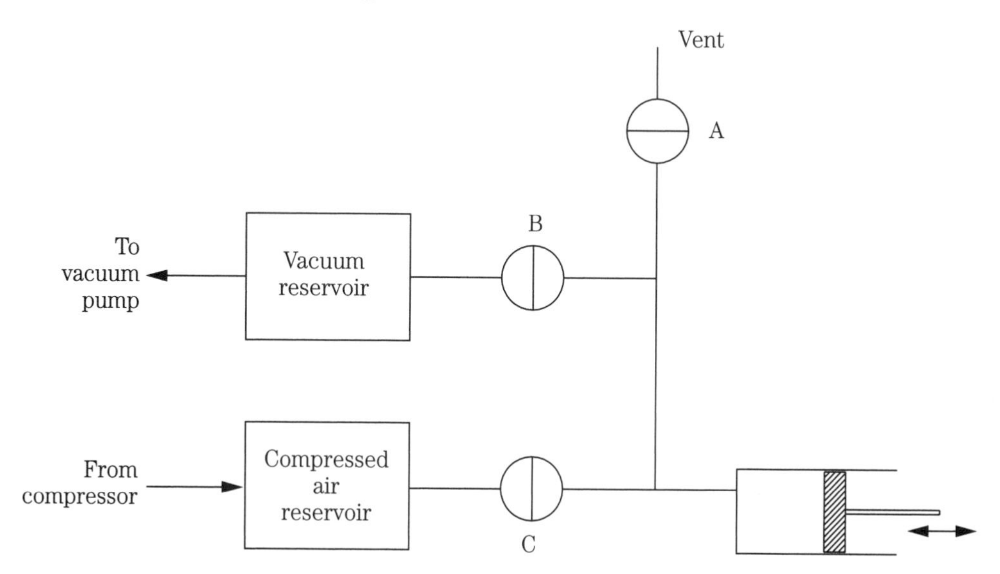

10-6 Schematic diagram of a hybrid position system that uses compressed air to extend the piston and vacuum to retract the piston.

To extend the piston, valves **A** and **B** are closed and valve **C** is opened to admit compressed air into the cylinder. Once the piston is at the desired extension, valve **C** is closed to hold the pressure inside the cylinder. Retracting the piston requires two steps. First, the pressure in the cylinder is vented by opening valve **A** with valves **B** and **C** closed. Then valve **A** is closed and valve **B** is opened to draw a vacuum in the cylinder.

Diaphragm actuators

In many applications it would not be practical to construct a vacuum piston that would be large enough to develop the required force. An alternative that is widely used in automotive applications employs a diaphragm that is moved by drawing a vacuum on one side while the other side is exposed to ambient pressure. A cross-section view of a typical diaphragm actuator used to operate an exhaust gas recycling (EGR) valve is shown in Fig. 10-7.

Vacuum connected to the vacuum inlet nipple lowers the pressure in the upper chamber causing the ambient pressure in the lower chamber to push the diaphragm upward. (The diaphragm is often constructed from thin metal and includes folds to allow it to flex.) The diaphragm pulls the actuator rod and attached valve plug up-

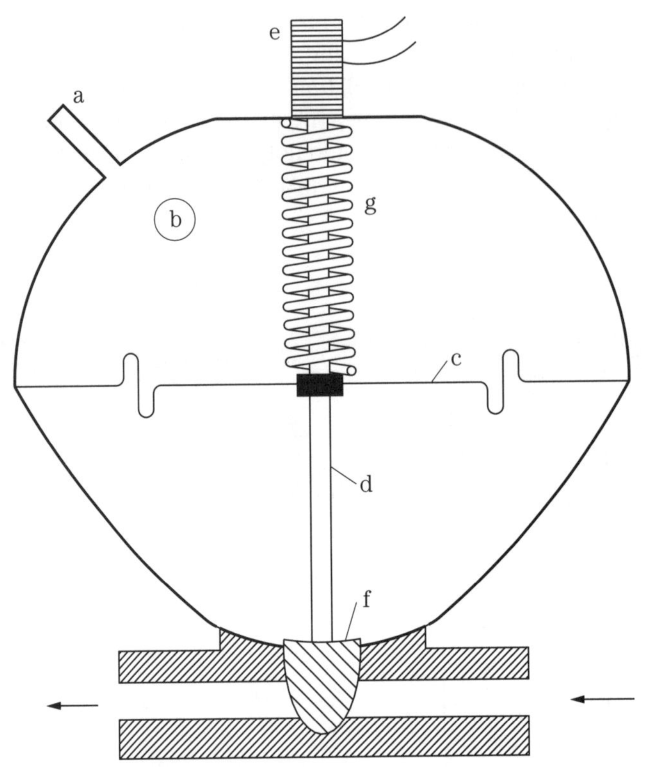

10-7 Diaphragm actuator attached to an EGR valve.

ward thus opening the valve and allowing gases to flow through the valve. The position of the valve is sensed by the valve position sensor, which is connected to the engine control computer. When the vacuum is released, the spring pushes the diaphragm down, causing the valve to close.

A diaphragm actuator with a metal diaphragm would be difficult to fabricate in the home shop. Figure 10-8 shows a cross-section view of a diaphragm actuator designed for home-shop fabrication. The chambers are made from plastic jars, and the diaphragm is cut from a sheet of rubber.

Vacuum gauges

Because most modern cars make extensive use of vacuum components in their emission-control systems, good dial type vacuum gauges are available at virtually any automotive supply store. Figure 10-9 is a photograph of a combination vacuum and pressure gauge sold by Sears. Assorted sizes of vacuum tubing and solenoid valves for vacuum lines are also available from automotive supply stores.

Homemade vacuum gauges

The use of homemade manometers as pressure gauges was discussed in chapter 9. Manometers are just as useful as vacuum gauges. However there is one major dif-

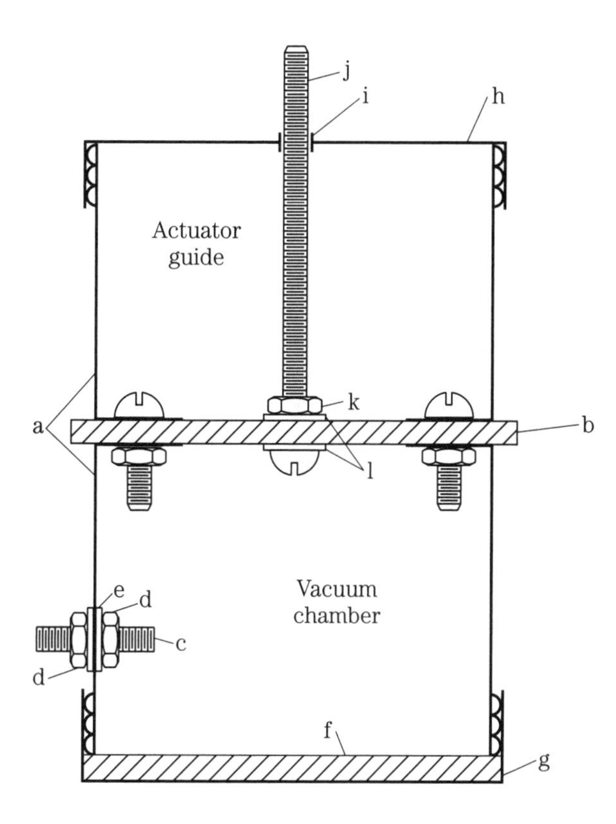

10-8
Homemade vacuum actuator made from plastic jars: (a) plastic jars, (b) rubber diaphragm, (c) threaded nipple, (d) nuts, (e) gasket washers, (f) jar-lid gasket, (g) solid jar lid, (h) jar lid with hole for actuator rod, (i) plastic bushing, (j) actuator made from long stove bolt, (k) nut securing actuator to diaphragm, (l) washers.

10-9
Combination vacuum and pressure gauge.

ference. Overpressure in a pneumatic system can blow the fluid out of the manometer, and if the fluid is water there may be no special precautions taken to capture the fluid in the event of a blow-out. On the other hand, manometers in vacuum systems are prone to having their contents sucked out rather than blown out. Many years ago, I had a manometer made from flexible tubing and filled with mercury. I had scavenged a compressor from a discarded refrigerator and I was experimenting with the compressor and manometer in the back yard. I was using the inlet of the compressor as a crude vacuum pump to collapse empty turpentine cans. I don't know if the vacuum was too strong, or if the manometer got tilted away from vertical, but suddenly all the mercury got sucked out of the manometer into the compressor. The compressor seized up and never ran again. Because of the nasty things that it does to the environment, mercury is not as easy to get as it was back then, but I imagine that water could also do serious damage to a compressor designed for Freon. Moral of the story: Install a liquid trap before the inlet of a vacuum pump and make sure the trap is large enough to hold the entire contents of any manometer that is connected to the system. Figure 10-10 shows the basic ideas involved in designing a simple trap.

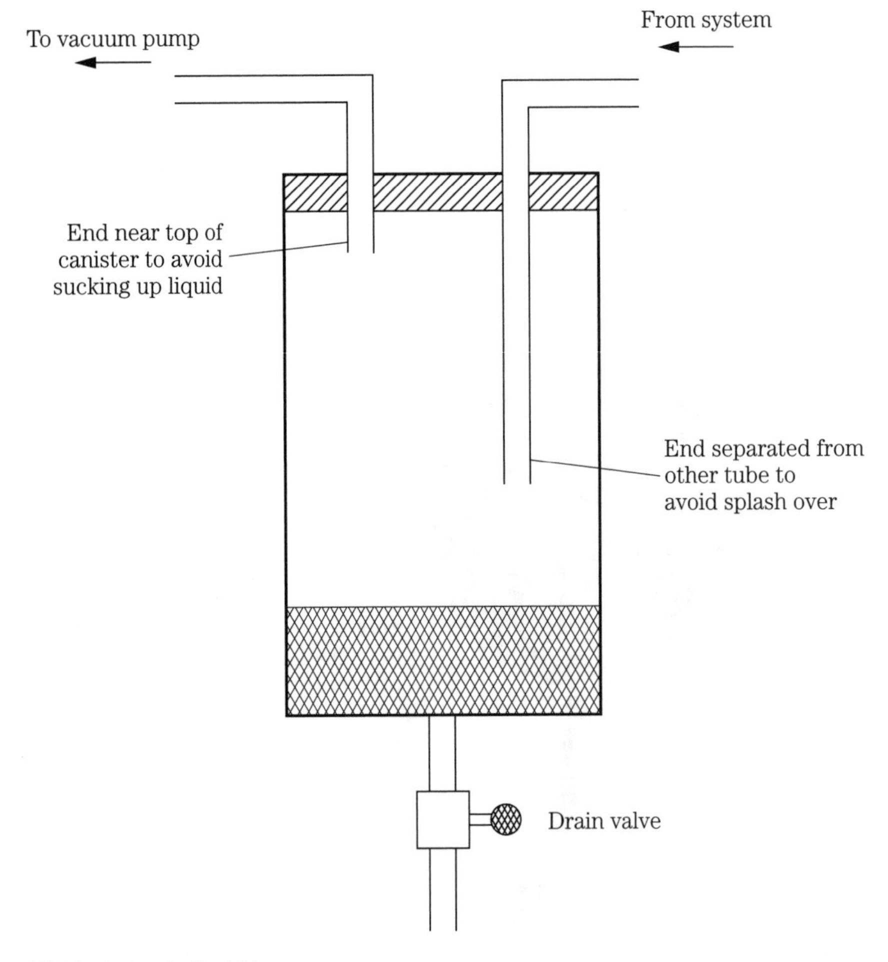

10-10 A simple liquid trap.

11
CHAPTER
Hydraulic systems

Hydraulic systems can be messy to work with and messy to operate in a home environment. However, for a given size, hydraulic force components provide greater force than their electromechanical or pneumatic counterparts. Cylinders and pistons are the hydraulic components most likely to be included in the design of a homebrew device because they are the most easily fabricated of the hydraulic force components.

Cylinders and pistons

Consider the cylinder and piston shown in Fig 11-1. The space within the cylinder is filled with liquid. A force of F is applied to the end of the piston as shown. Unlike gases, liquids are for all practical purposes incompressible. Therefore the piston will not move unless some of the fluid is allowed to escape from the cylinder. The piston will exert a pressure on the liquid, and this pressure will be distributed throughout every part of the liquid. If A denotes the area of the piston face that is exposed to the liquid, the pressure in the liquid will be F/A. This relationship works both ways. If the pressure within the liquid is P, the force exerted against the inside face of the piston will be P × A.

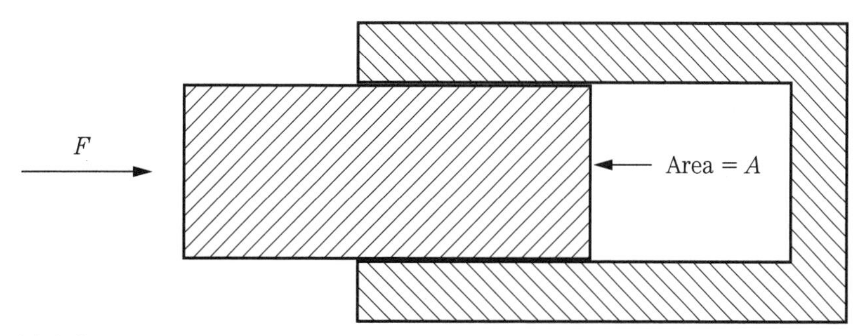

11-1 Cutaway view of a cylinder and piston.

Hydraulic press

Consider the *hydraulic press* shown in Fig. 11-2. The force applied to piston **A** creates a pressure of $P = F_1/A_1$ in the liquid. This pressure is transmitted to every part of the liquid, including the liquid in cylinder **B**. This pressure will exert a force F_2 on piston **B** that is equal to $P \times A_2$. The relationship between F_1 and F_2 can be established as:

$$\frac{F_1}{A_1} = P = \frac{F_2}{A_2} \qquad (11\text{-}1)$$

By selecting cylinder sizes so that $A_2 > A_1$, the system can be used to multiply force.

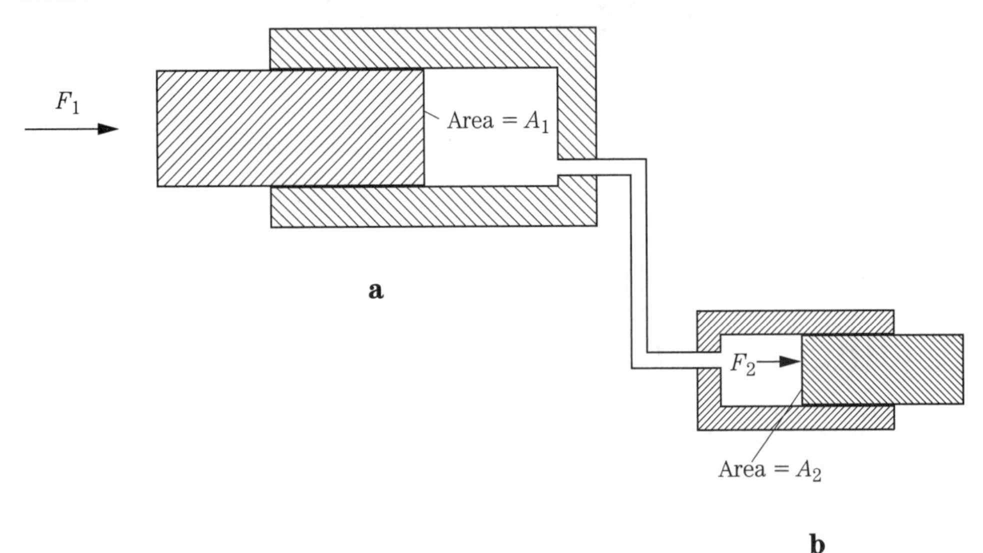

11-2 Cutaway view of the cylinders and pistons in a hydraulic press.

Example 11-1

For the hydraulic press depicted in Fig. 11-2, assume that A_1 = 6 sq. in., and F_1 = 2 lb. Determine the value of A_2 needed to make F_2 = 20 lb.

$$A_2 = \frac{A_1 F_2}{F_1} = \frac{(6)(20)}{2} = 60 \text{ in}^2$$

Unlike the closed cylinder of Fig. 11-1, the pistons in the system of Fig. 11-2 can move. If piston **A** moves into its cylinder, it will displace fluid through the tubing and into cylinder **B**. This additional fluid will cause piston **B** to move farther out of its cylinder. When piston **A** moves inward a distance of X_1 the volume of fluid displaced will be A_1X_1. This amount of additional fluid introduced into cylinder **B** will cause piston **B** to move out a distance of:

$$X_2 = \frac{A_1X_1}{A_2} \tag{11-2}$$

In the terminology of simple machines, we can say that the theoretical mechanical advantage (TMA) of the hydraulic press is given by:

$$\text{TMA} = \frac{A_2}{A_1} = \frac{F_2}{F_1} = \frac{X_1}{X_2}$$

The amount of work done on piston **A** is F_1X_1 and the amount of work done by piston **B** is F_2X_2. Substituting for F_2 and X_2 the expressions given by Eqs. 11-1 and 11-2 we obtain:

$$F_2X_2 = \left(\frac{F_1A_2}{A_1}\right) \times \left(\frac{A_1X_1}{A_2}\right) = F_1X_1$$

So it turns out that the work done by piston **B** equals the work done on piston **A**. Of course, in a real hydraulic system, there will be some losses such as friction that will keep the system from being 100 percent efficient, but nevertheless, the efficiency of hydraulic systems is usually very high.

The pistons in Figs. 11-1 and 11-2 were depicted in a horizontal position. In any other position, the weight of the piston itself will exert some force that will be directed either toward or away from the fluid in the cylinder. The direction of the force will depend on the orientation of the cylinder. (See Fig. 11-3.) When a cylinder is on a slant between horizontal and vertical, the weight of the piston should be resolved into components that are parallel and perpendicular to the piston's direction of motion. The parallel component can then be added to or subtracted from the force on the piston.

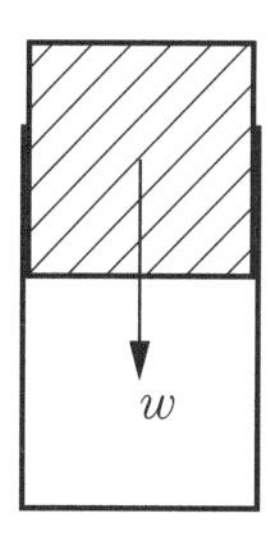

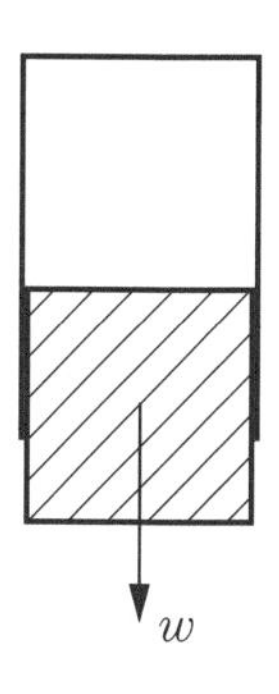

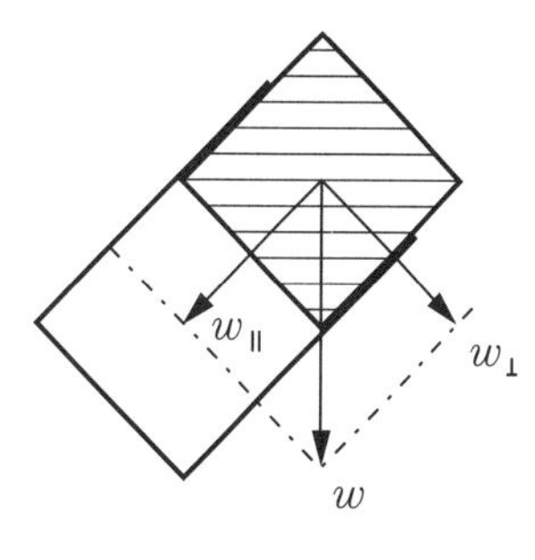

11-3 Three different orientations of a cylinder showing how the weight of the piston can exert a force against the fluid within the cylinder.

Normally the force of the hydraulic fluid in a cylinder moves the piston in only the "outward" direction. Simply removing the pressure and allowing the hydraulic fluid to escape from the cylinder will not be sufficient to move the piston back into the cylinder. Some means must be provided to cause return of the piston. Springs are frequently used for this purpose. Both the gripper shown in Fig. 11-4 and the

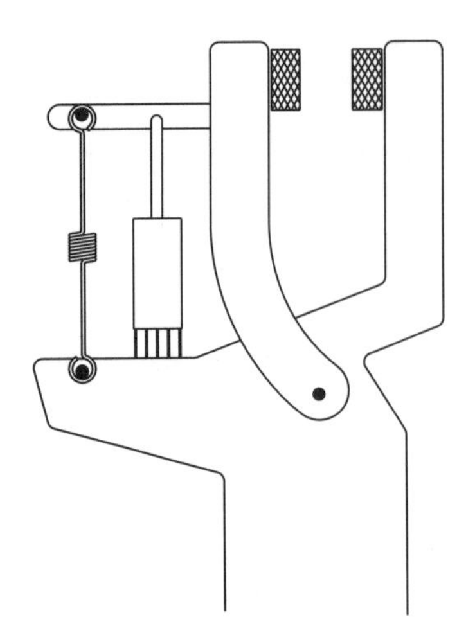

11-4
Hydraulic gripper with spring return.

brake shoe mechanism shown in Fig. 11-5 use springs to return the piston when the hydraulic pressure is relaxed.

The double-acting cylinder depicted in Fig. 11-6 is an alternative to a spring return system. In this type of cylinder, admitting pressurized fluid into the left end of the cylinder will push the piston to the right. On the other hand, relaxing the pressure on the left end and admitting pressurized fluid into the right end of the cylinder will push the piston to the left.

Applications for the hydraulic press

How can a hydraulic press be integrated into an electromechanical system? One way is to use a motor to apply force to the piston as shown in Fig. 11-7. The motor is controlled electrically through its power leads. The motor drives the gearbox, which increases the torque so that the threaded rod can be turned. The threaded rod mates with a threaded hole in the follower block. If the motor rotates in one direction, the follower will move away from the gearbox, forcing the piston into the cylinder. This will force hydraulic fluid out of the tube which is connected to the output cylinder. If the motor rotates in the other direction, the follower will move towards the gearbox, pulling the piston out of the cylinder. This seems like a lot of trouble to go to; if we have to use a motor anyway, why not have the motor perform the desired task directly without having to use the hydraulic press as an intermediary? Well, there are several good reasons why we would want to use the apparently more complicated hydraulic approach.

Hydraulic robot gripper

Consider the design of a robot that needs a small but powerful gripper. A motor with sufficient power could be too big to fit within the application's size limitations. A hydraulic press could provide the solution. Place a small cylinder and piston on the manipulator as needed to operate the gripper. Connect this cylinder via tubing to a piston located inside the robot's body. The piston can then be driven by a large mo-

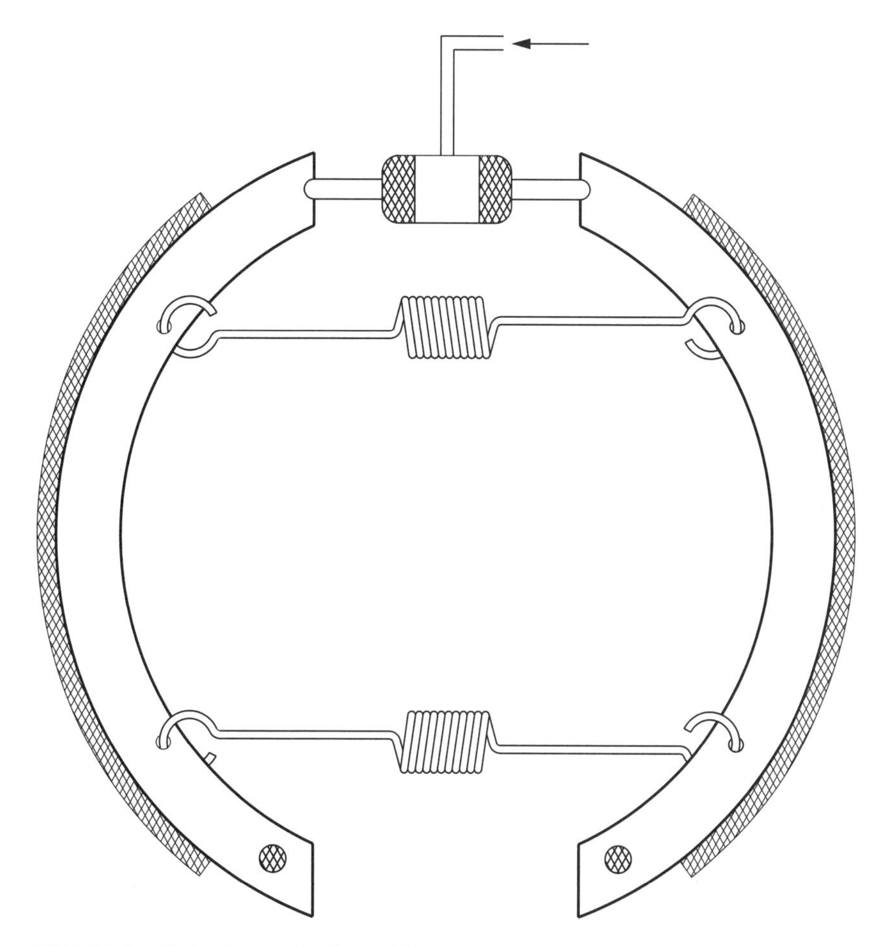

11-5 Hydraulic brake mechanism with spring return.

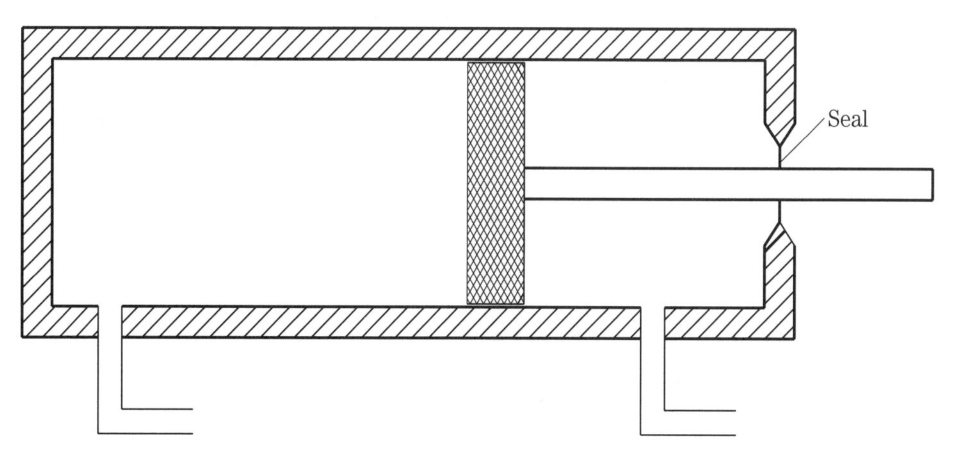

11-6 Cutaway view of a double-acting cylinder.

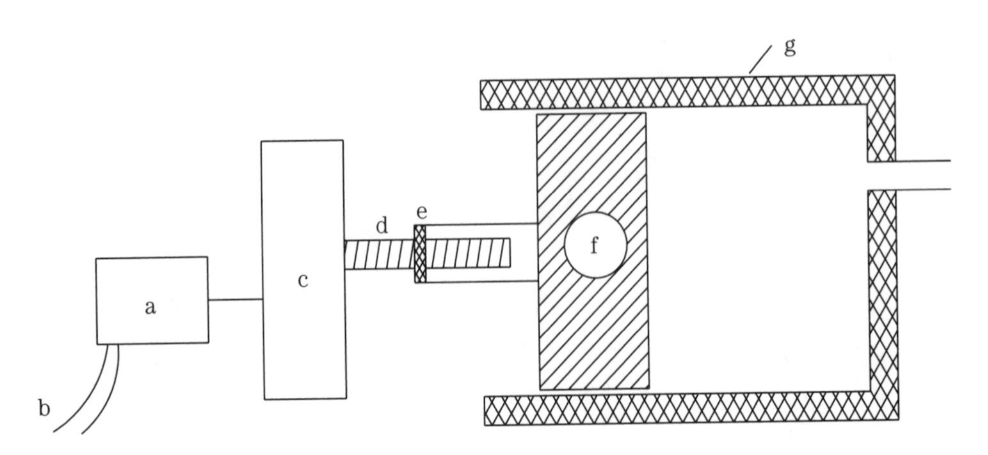

11-7 Using an electric motor to drive a piston. The motor (a) is controlled electrically via its power leads (b). The motor drives gear box (c) which increases the torque so that the threaded rod (d) can be turned. The threaded rod mates with a threaded hole in the follower block (e). If the motor rotates in one direction, the follower will move away from the gearbox forcing the piston (f) into the cylinder (g). This will force hydraulic fluid out of the cylinder into the tube which is connected to the output cylinder.

tor that is powerful enough to get the job done. The output piston can be rather small but still be capable of closing a gripper with considerable force. The force exerted by a piston is not directly limited by its size. The force exerted by the piston, as we saw earlier in this chapter, is equal to the pressure of the hydraulic fluid times the end-area of the piston. We can compensate for a small piston area by using a high fluid pressure. Of course, we must be careful to ensure that the pressure used does not exceed the pressure ratings of the cylinder, tubing, or any other components of the hydraulic system.

Hydraulic cylinders

The wheel cylinder in most drum brake assemblies is double-ended. As shown in Fig. 11-8, this consists of an open-ended cylinder with two pistons. The hydraulic fluid is introduced between the two pistons, causing both pistons to exert approximately the same force and move approximately the same travel distance. Notice that there is nothing in the wheel cylinder itself that prevents the hydraulic fluid from pushing the pistons completely out of the cylinder. Only the limited travel of the brake shoes prevents this from happening in actual use. Anyone using automotive wheel cylinders in other applications should keep this fact in mind and be sure to provide stops on the driven element so that the pistons are kept inside the cylinder.

Pistons with O-ring seals as described in chapter 9 can also be used as hydraulic cylinders. Just be certain that the rubber used in the O-ring seal is not dissolved or attacked by the hydraulic fluid.

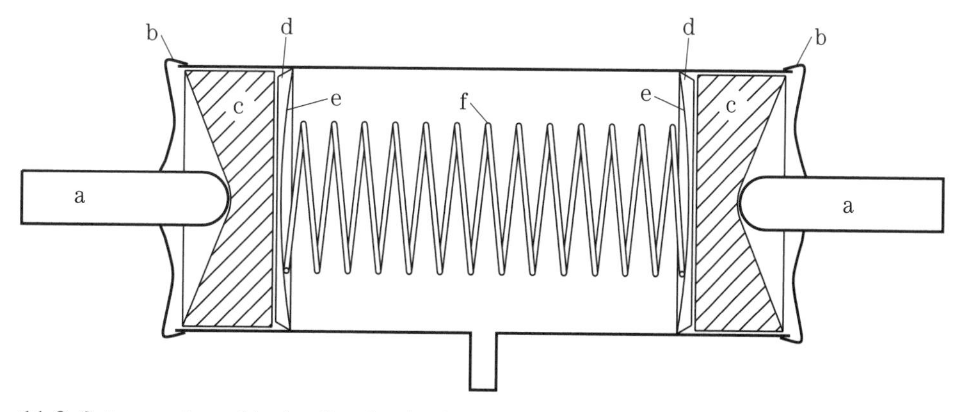

11-8 Cutaway view of hydraulic wheel cylinder: (a) external linkage, (b) rubber end boots, (c) metal pistons, (d) rubber cups, (e) expanders [to force cup edges against cylinder walls], (f) spring, (g) fluid inlet.

Hydraulic pumps

We can modify the hydraulic press shown in Fig. 11-2 by removing the input cylinder and replacing it with a hydraulic pump as shown in Fig. 11-9. This is the sort of arrangement used to operate the hydraulic pistons that raise and lower snow plows on some trucks. The pump can have its own electric motor, or it can be operated by a belt driven by the vehicle engine. Stores and mail-order companies that sell truck accessories are the best sources of hydraulic pumps for amateurs.

Hydraulic accumulators

Hydraulic *accumulators* are devices for storing hydraulic fluid under pressure. Several different types of accumulators are depicted in Figs. 11-10 through 11-15. For some applications, the gas-loaded piston accumulator can be used in systems not having a hydraulic pump. The fluid can be admitted to the cylinder under low pressure conditions and then pressurized using a pneumatic compressor.

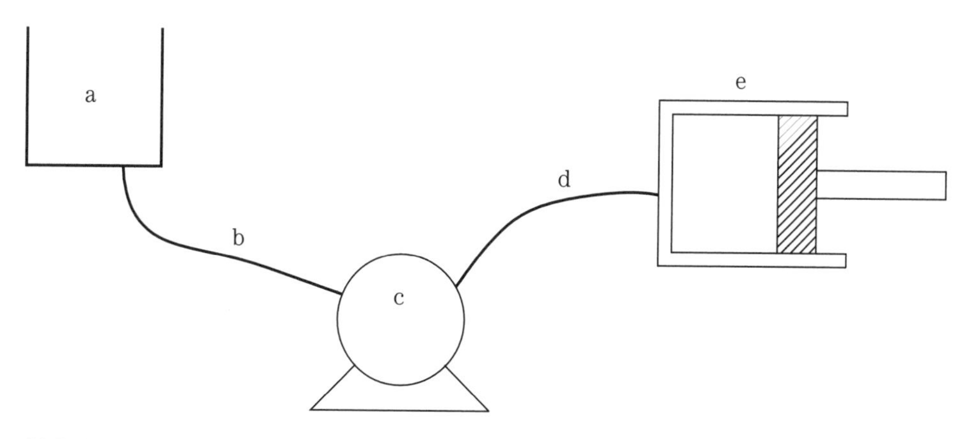

11-9 A system for pump operation of a hydraulic cylinder. The components are: (a) fluid return, (b) pump input line, (c) hydraulic pump, (d) pump output line, and (e) cylinder assembly.

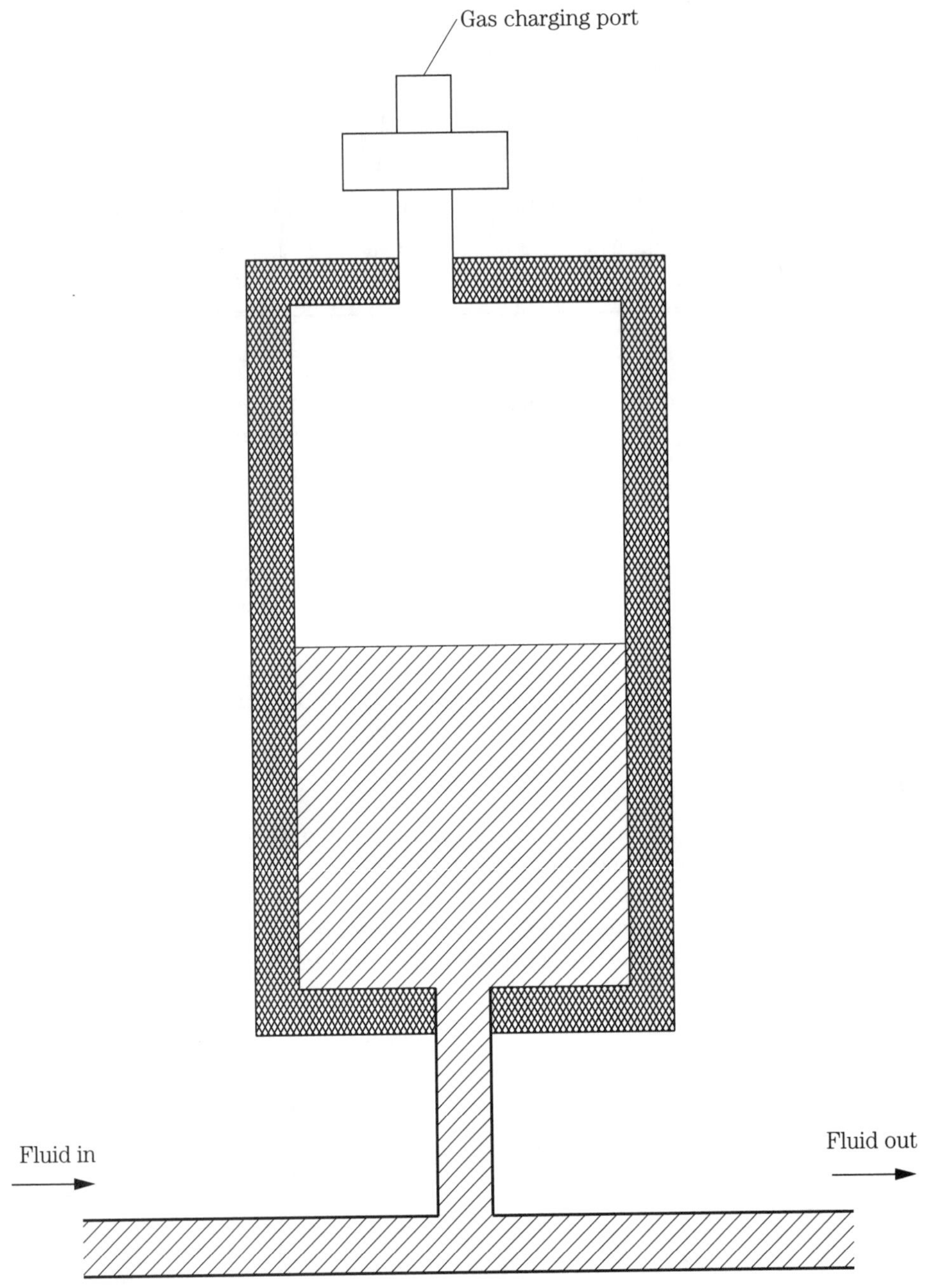

11-10 Nonseparator gas loaded accumulator.

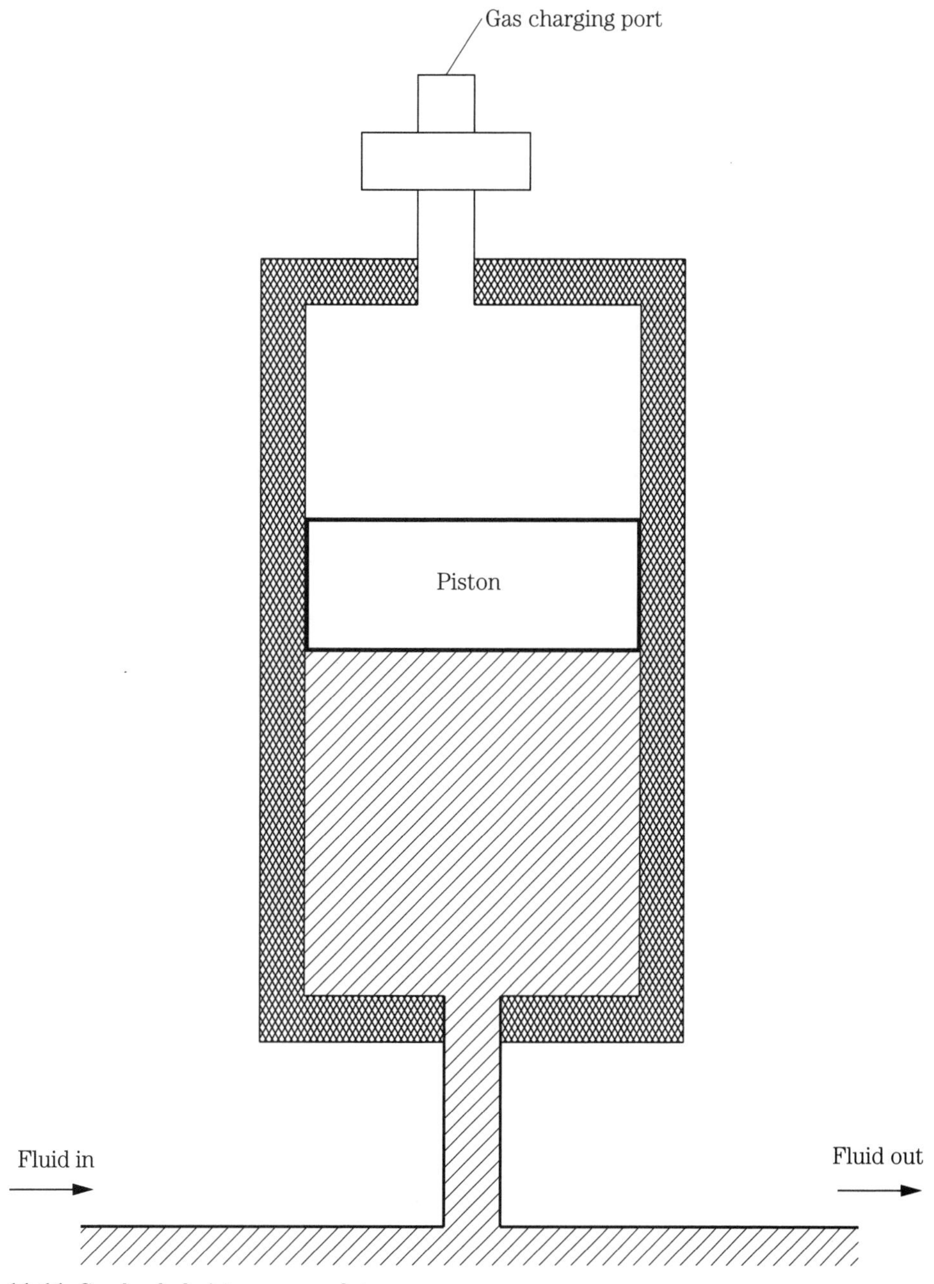

11-11 Gas-loaded piston accumulator.

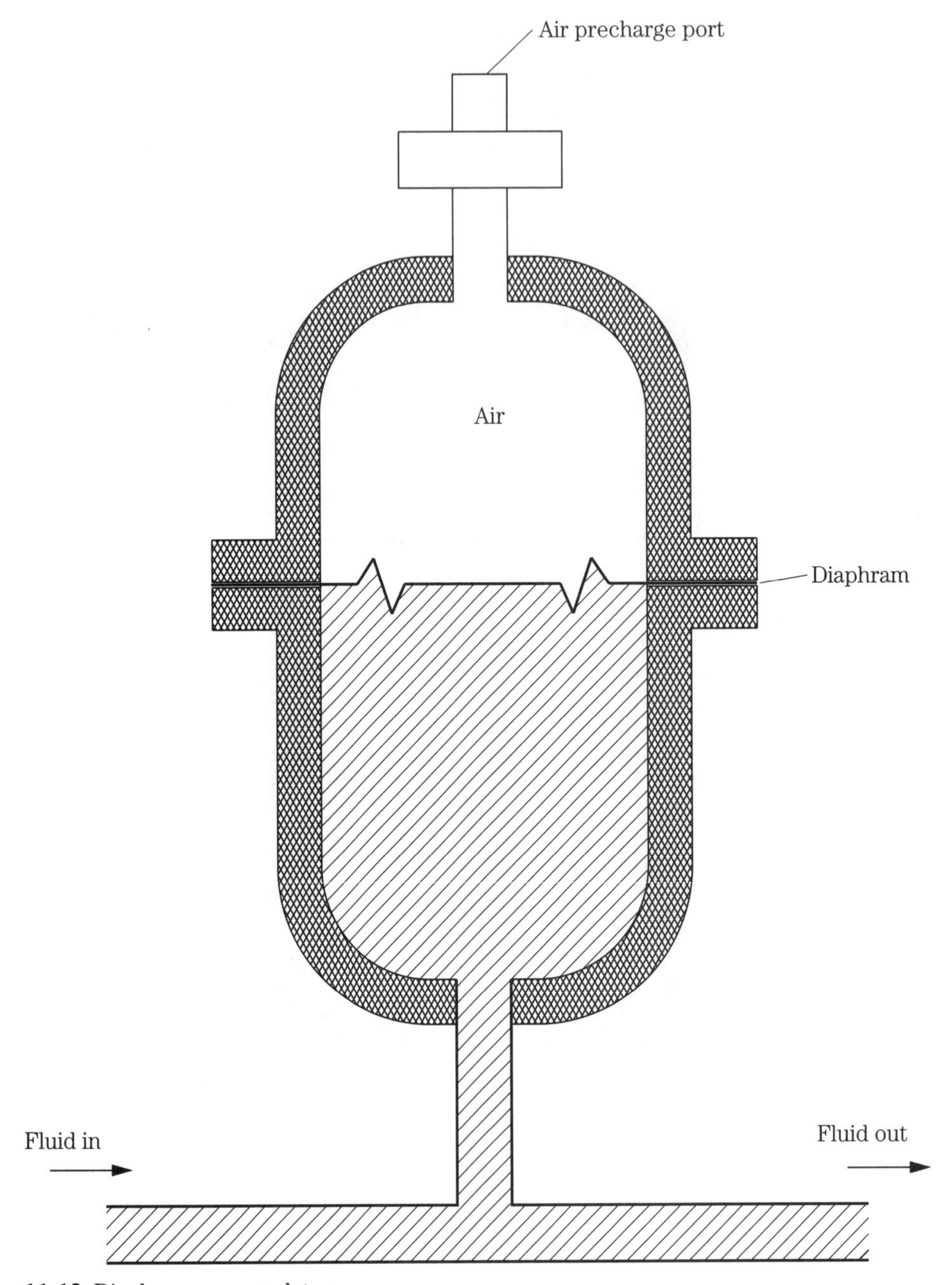

11-12 Diaphragm accumulator.

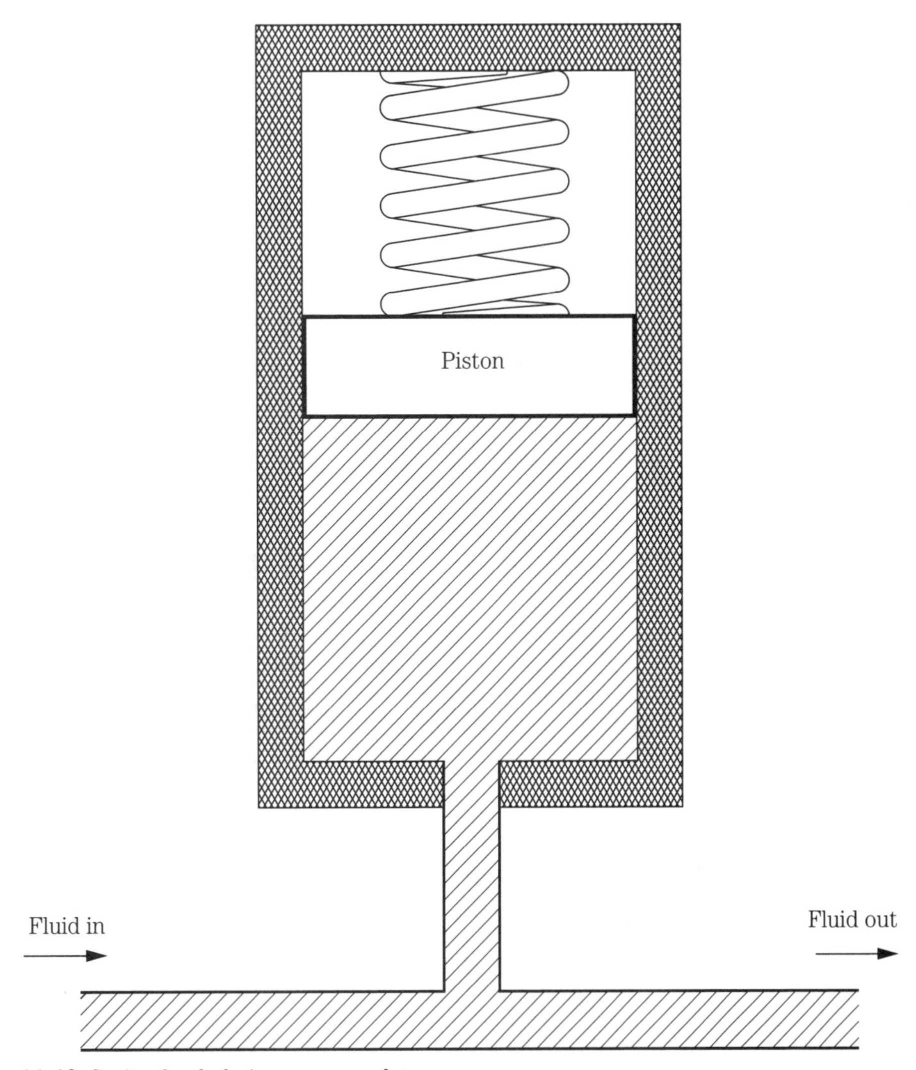

11-13 Spring-loaded piston accumulator.

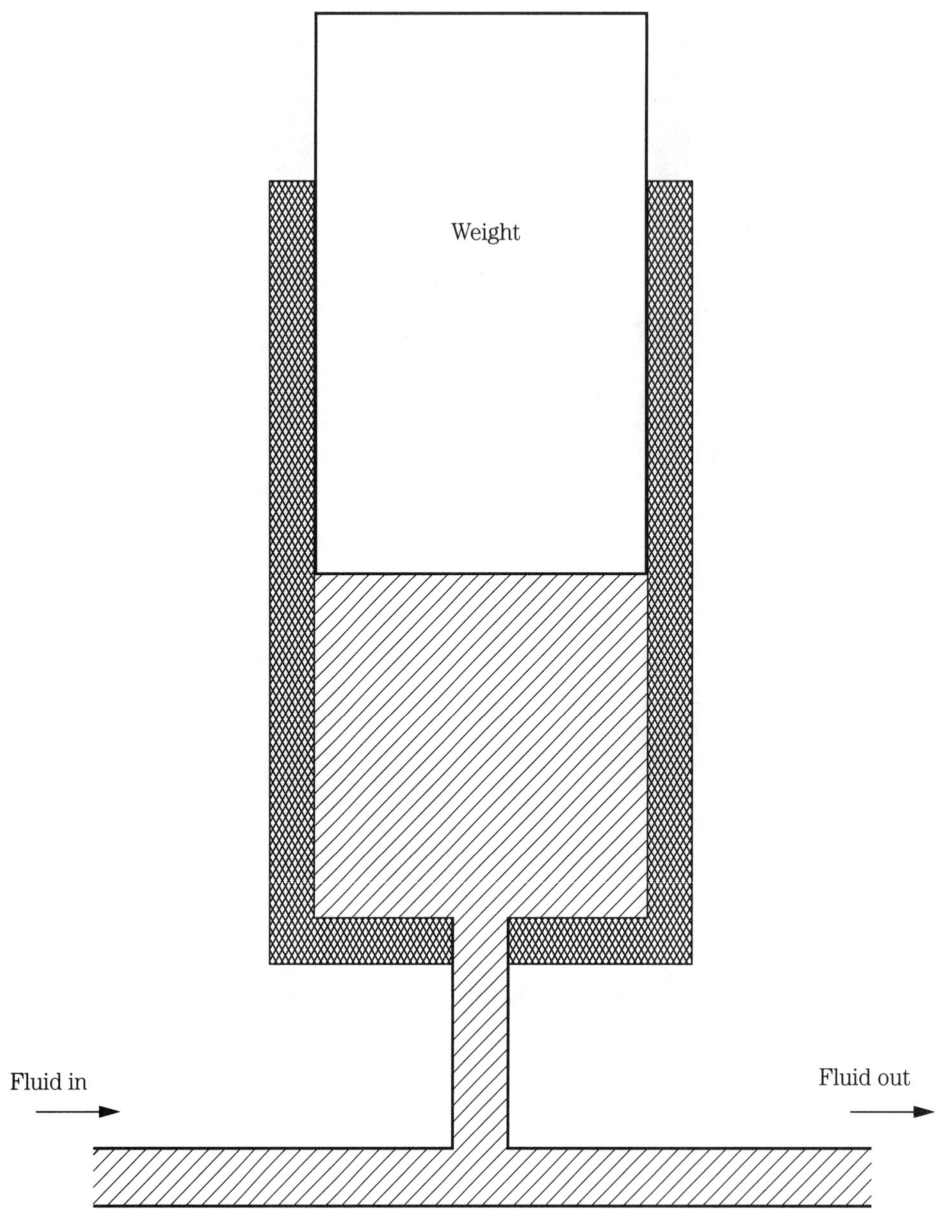

11-14 Weight-loaded piston accumulator.

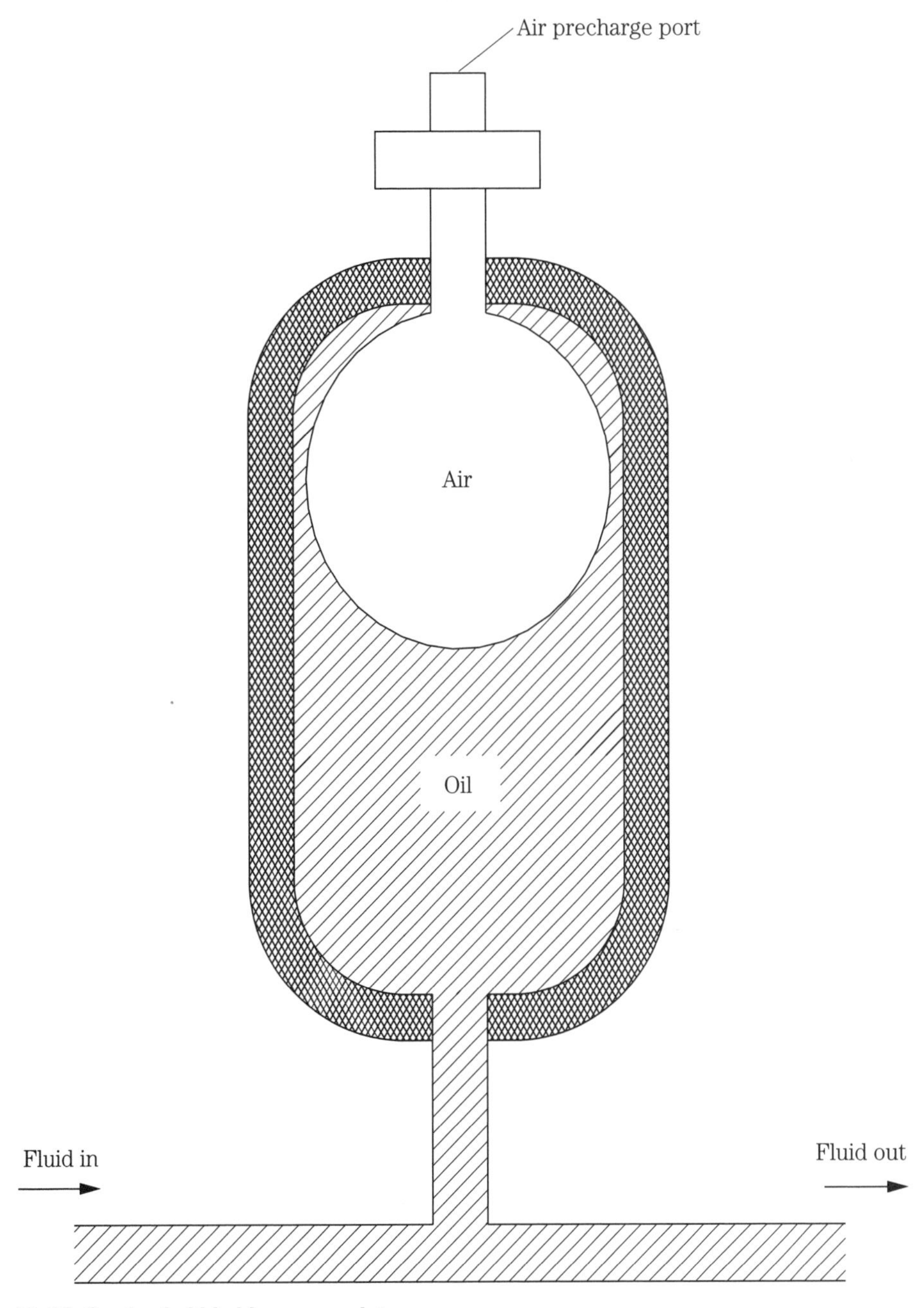

11-15 Gas-loaded bladder accumulator.

12
CHAPTER
Wheeled vehicles

A wheeled robot is, among other things, a wheeled vehicle. Much of the technology developed by vehicle designers can be applied directly to the problems of robot mobility, but there are many design choices that need to be made. One very fundamental question is, "How many wheels should a robot have?" The right answer depends on the intended application:

- An interesting project in control system design might be a robotic unicycle or bicycle like the ones depicted in Fig. 12-1, in which a control system moves weighted arms in or out as needed to maintain the vehicle's balance. Although potential science fair projects in their own right, these vehicles would not be a good platform on which to construct additional robot capabilities.
- Three wheels is the minimum number needed to have unconditional stability on a level surface. Many successful three-wheeled robots have been designed and constructed.
- Based on experiences with cars, wagons, and tool carts, our first instinct is usually to think in terms of four-wheeled vehicles with steering in the front.
- Many all-terrain vehicles have six wheels, and by analogy, six might be the right number of wheels for a robot that will be used outdoors on unpaved surfaces.
- Some military and construction vehicles designed for extremely rugged off-road applications use treads in place of wheels.

After the number and configuration of the wheels is decided on, the design of a wheeled platform involves a number of additional mechanical issues that must be confronted by the experimenter. This chapter will explore some of these issues with an emphasis on robotic applications.

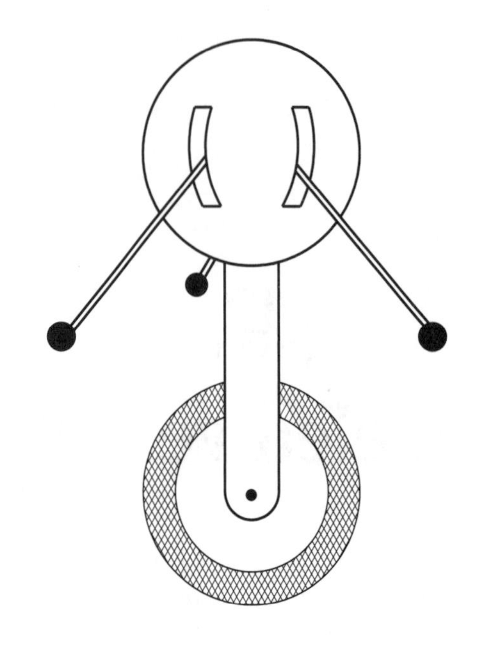

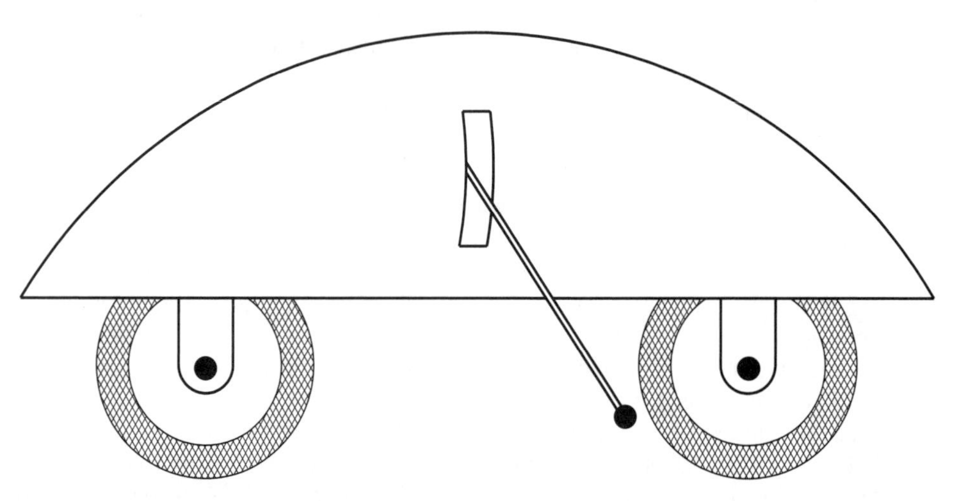

12-1 Unicycle and bicycle robots.

Four-wheeled vehicles

Although common in the everyday world, four-wheeled vehicles come with a host of problems. Everyone has had an experience with a four-legged table that wobbled because only three of the legs were in contact with the ground. A four-wheeled vehicle is subject to the same sort of situation in which only three of the wheels are in contact with the ground. Remedies can be quite complicated; this is why so much of an automobile chassis is dedicated to suspension and steering. A schematic diagram of a typical automotive wheel suspension scheme is shown in Fig. 12-2. A wheel mounted in this way has some limited freedom to move up and down so that it can follow the ups and downs of an imperfect road surface. Good suspensions take up space. In a car, most of the space under the fenders on either side of the trunk and engine compartment is taken up by suspension components. Design and construction of an independent wheel suspension system that is suitably compact for robotic applications could present a real challenge.

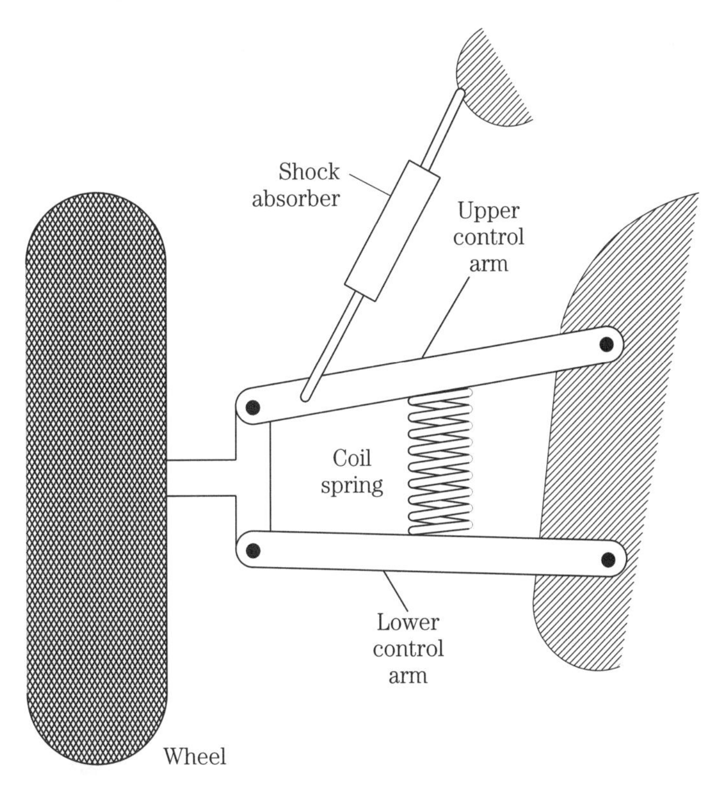

12-2 Typical scheme for independent suspension of automobile wheels.

Consider a vehicle that is going around a curve with its wheels following arced paths as shown in Fig. 12-3. The wheels on the outside of the curve must travel a greater distance than the wheels on the inside of the curve. Since the wheels all have the same circumference, this means that the wheels on the outside of the curve must turn faster than the wheels on the inside of the curve. Most vehicles have a single engine with a single transmission shaft. This is a problem for automotive designers who must simultaneously obtain two different wheel speeds from a single transmission shaft. In a car with rear-wheel drive, the two different wheel speeds are obtained via the *rear differential*. In a car with front-wheel drive, the two different wheel speeds are obtained from a differential built into the output of the transaxle.

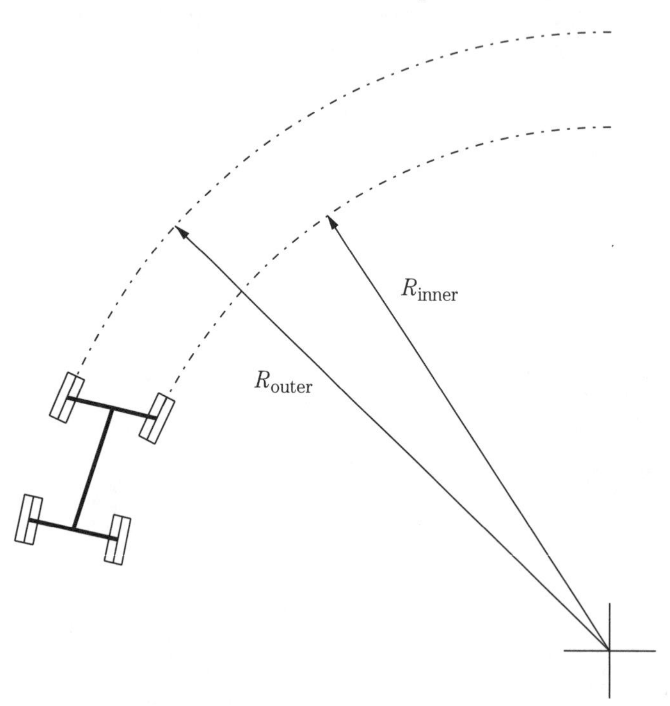

12-3 Paths followed by the wheels of a robot negotiating a curve.

Using a differential in a robot drive is a nontrivial matter, but it can be done. The workings of a rear differential are depicted in Fig. 12-4. Sometimes, amateur robot builders (and even professional designers of cheap robot kits) ignore the need for different speeds and wind up with a robot that runs okay on the straightaways but maybe not so smooth on curves. If the drive wheels on both sides are forced to run at the same speed, the robot will compensate on the curves with some amount of slipping of the faster wheel and some amount of dragging of the slower wheel. There **is** a better solution!

The drive wheels on each side of a robot can be driven by a separate electric motor. Either a microprocessor or dedicated logic can be used to drive the two motors at different rates, with the difference between the rates depending on the turning radius being executed.

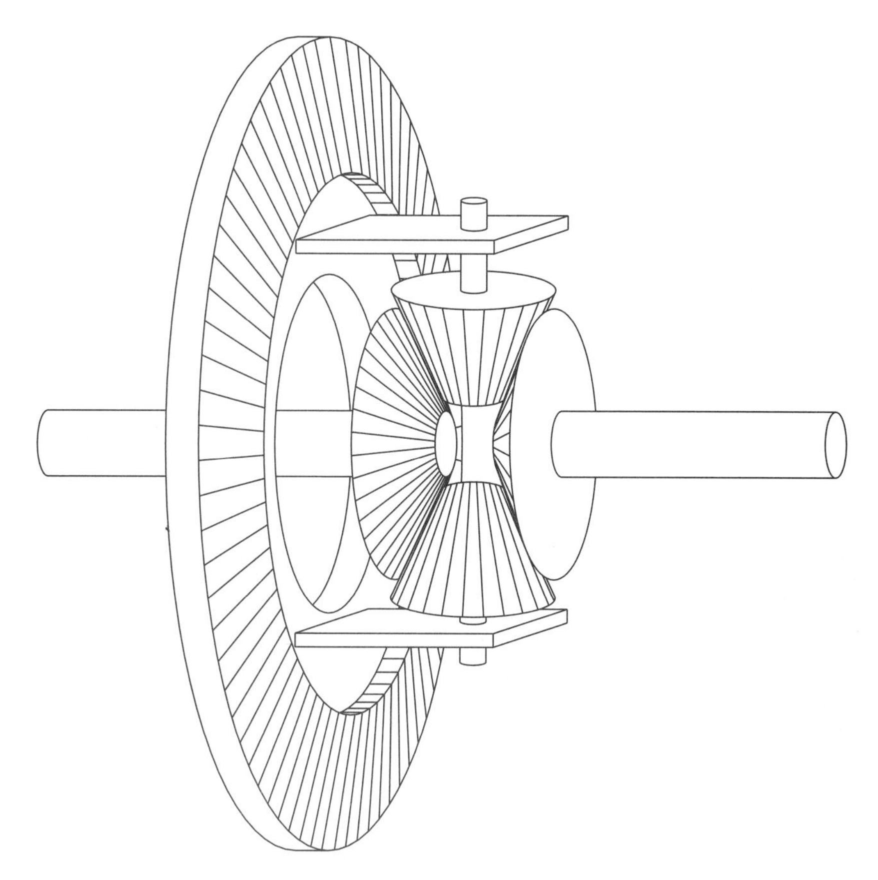

12-4 Typical design for differential gearing used in rear-wheel drive cars to allow drive wheels to turn at different rates while negotiating curves.

Steering

One approach to steering a robot is similar to the approach used in cars. As shown is Fig. 12-5, four links are joined on pivots so that they form a parallelogram. One long edge of this parallelogram is fixed to the chassis, and the wheels are attached to the short sides as shown. Moving the free long side to the left slants the wheels to the right. Conversely, moving the long side to the right slants the wheels to the left. The movement of the long side can be accomplished with a rack and pinion, or cable and capstan. When the wheels are slanted, the perpendicular distance between link **F** and link **A** is less than when the wheels are pointed directly forward. This means that if link **F** is fixed, link **A** will move fore and aft as well as left and right. If a rack is mounted on the side of link **A** as shown in Fig. 12-6, the fore-aft motion of link **A** will make it difficult to keep the pinion meshed with the rack.

If a wide rack is mounted on top of link **A** as shown in Fig. 12-7, the rack can move fore and aft under the pinion without losing mesh. This mechanism may exhibit some binding if everything is not carefully aligned or if the rack pitch is too fine.

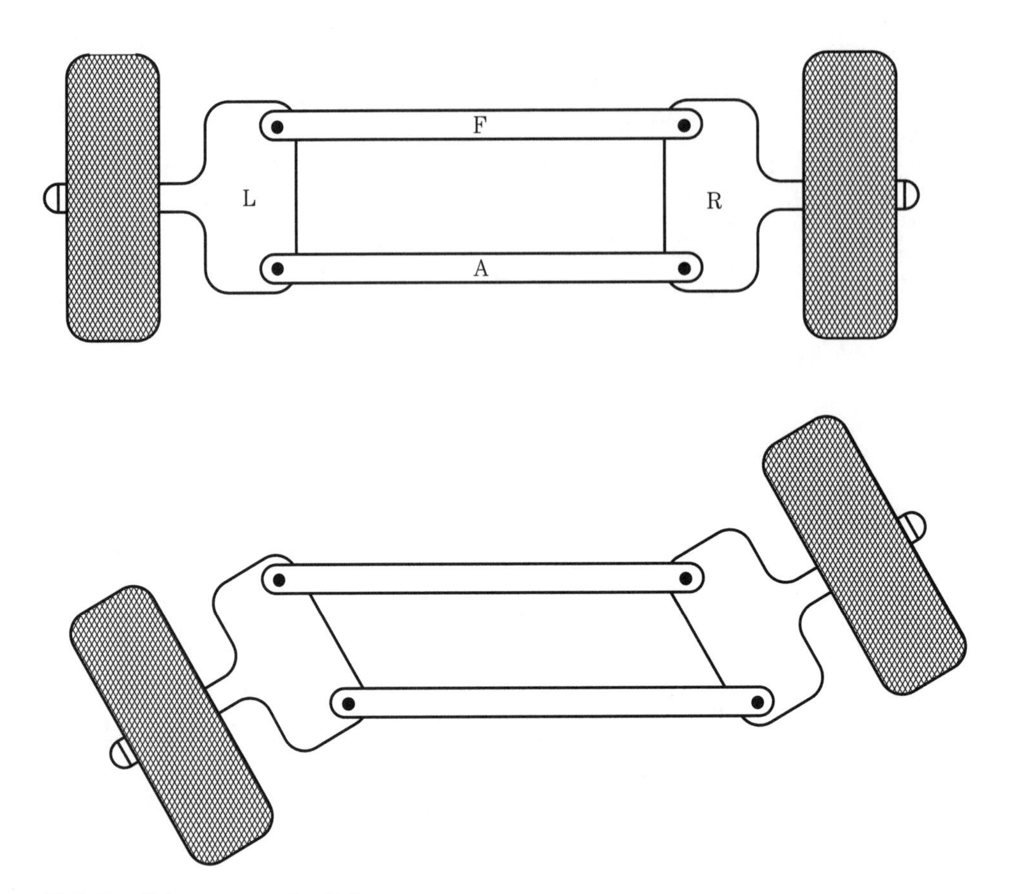

12-5 Parallelogram steering linkage.

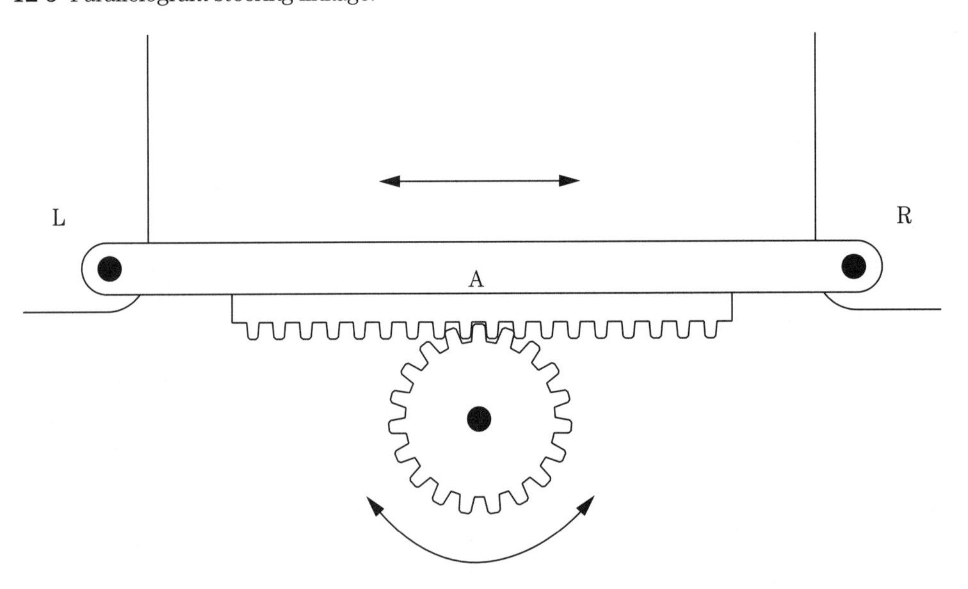

12-6 Closeup view of steering linkage showing rack mounted to side of link **A**.

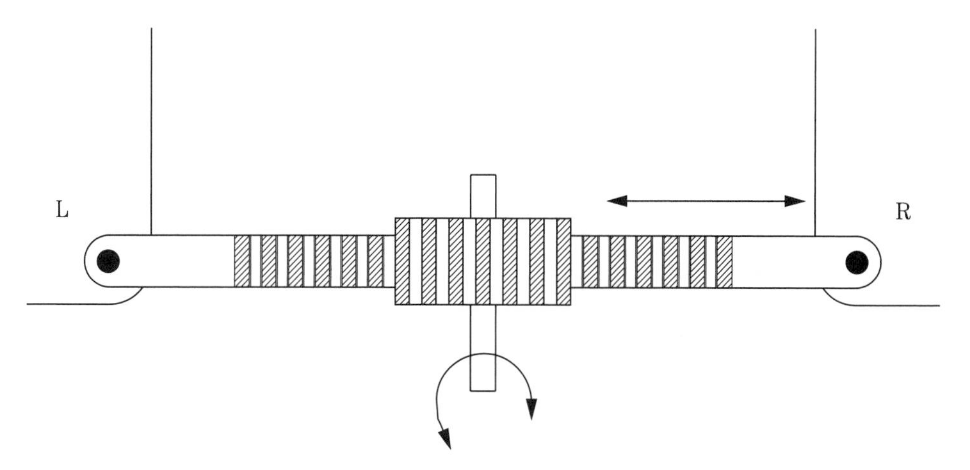

12-7 Closeup view of steering linkage showing rack mounted to top of link **A**.

The cable and capstan approach illustrated in Fig. 12-8 is more tolerant of imprecisions in fabrication and alignment.

A steering approach that has proven to be more popular with robot designers uses swivel-mounted wheels like those shown in Fig. 12-9. Hard rubber wheels of this sort are sold as *casters* for use on rolling tool cabinets favored by many gas station mechanics. These casters are available in hardware stores and home centers in sizes up to 5" in diameter.

For unpowered use, these casters can be used "as is." If we want to supply drive to a swivel-mounted wheel while maintaining the swivel operation, we must mount the drive motor so that it swivels with the wheel. There are ready-made motorized wheels available, but their popularity has driven their prices on the surplus market to around $50 per wheel. Depending on the availability of suitable motors and gears,

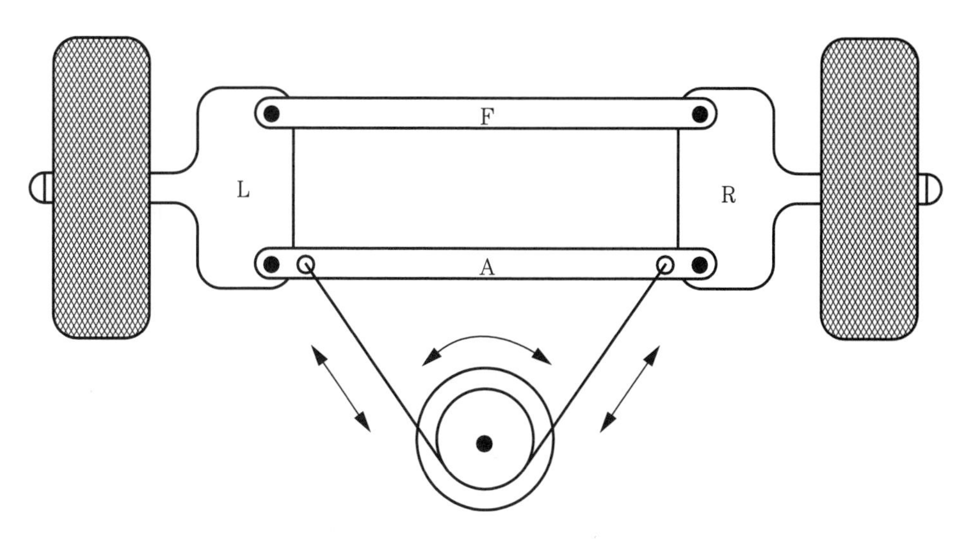

12-8 Steering linkage with cable and capstan drive.

12-9 Assortment of casters available from hardware stores.

it may be possible to make your own for significantly less. Figure 12-10 shows one way that a motor can be added to a caster so that the wheel and motor are both steered together as a unit.

Three-wheeled vehicles

There are really only two three-wheeled configurations that make sense. The cheap scheme places wheels at each vertex of an isosceles triangle as shown in Fig. 12-11. The base of the triangle corresponds to the rear of the vehicle and the vertex opposite the base corresponds to the front of the vehicle. The two wheels at the base vertices are fixed, and the wheel in front is steerable. I refer to this as the cheap scheme because the two rear wheels can be fixed on a common axle and powered by a single motor, with the front wheel steerable but unpowered.

From an overall stability viewpoint, it is probably best to make the two sides of the triangle approximately equal to the base. However, there is a range of usable triangle proportions. If the two equal sides of the triangle are made too large relative to the base, side-to-side stability will suffer, especially if the center of gravity is well forward in the vehicle. If the sides are made shorter than the base, front-to-back stability will suffer.

The deluxe three-wheeled scheme places a steerable motorized wheel at each vertex of an equilateral triangle. This scheme provides the ultimate in maneuverability. The vehicle can make a square corner by rolling forward, stopping, swiveling all wheels by 90 degrees, and rolling off in a new direction perpendicular to the old. A separate motor for each wheel makes possible some very elegant solutions to the differential drive problem inherent in negotiating curves.

Six-wheeled vehicles

Vehicles that use six or more wheels usually provide drive to all wheels, with all the wheels on each side running at a single rate. There isn't much advantage in driv-

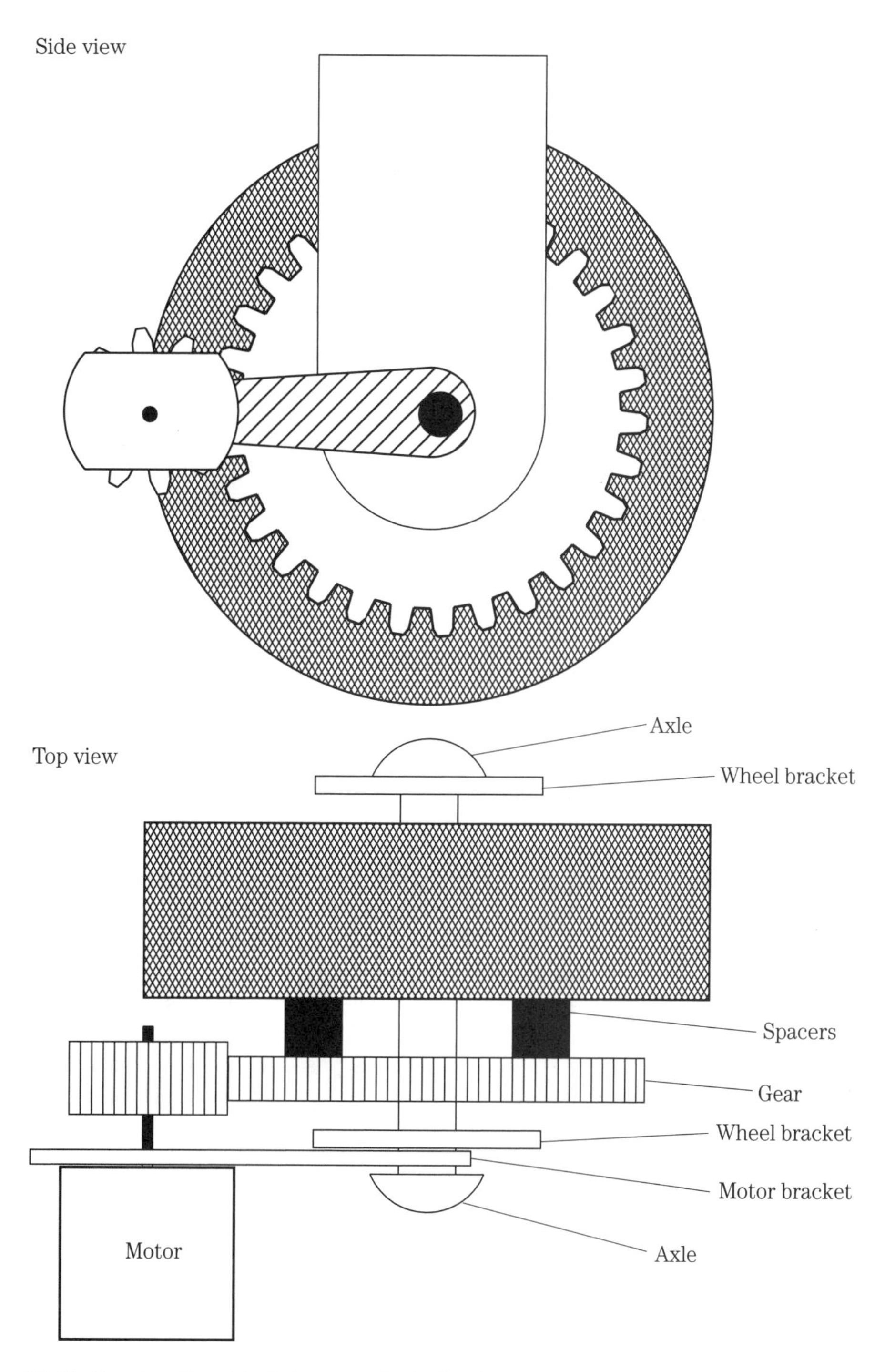

12-10 Construction details for motorized wheel.

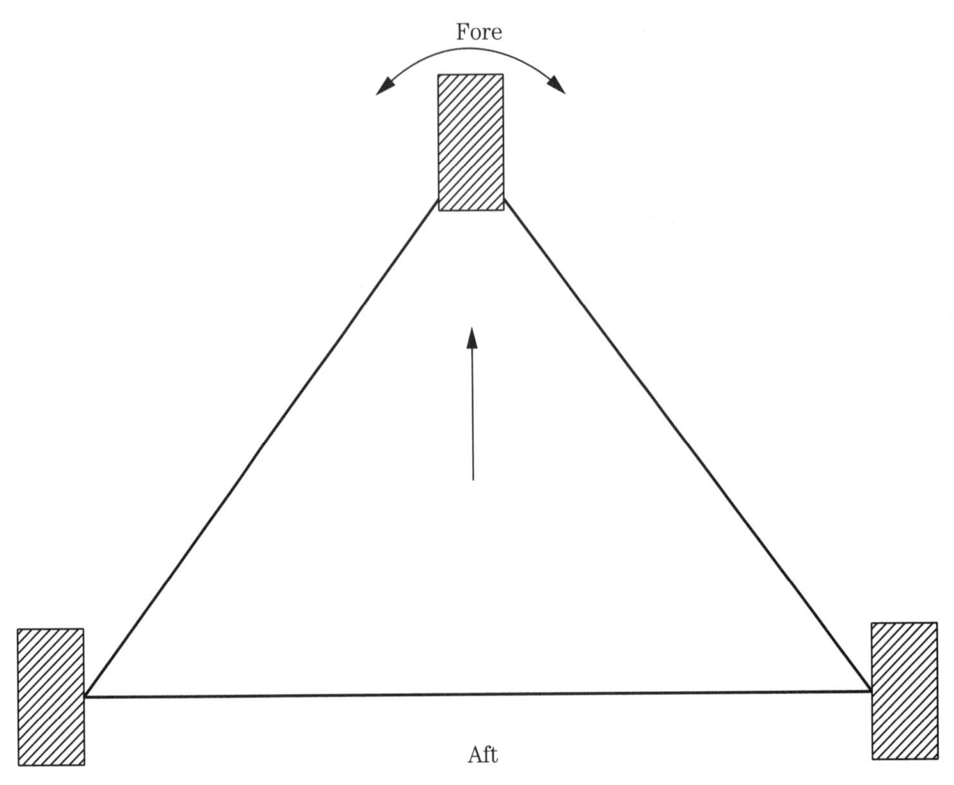

12-11 Three-wheeled configuration with two fixed wheels aft and one steerable wheel forward.

ing individual wheels at different rates, but there is an opportunity for added complexity and expense. However, there is an advantage in having left-side wheels independent of the right-side wheels. During turning maneuvers, the wheels on the outside of the turn can be made to run forward while the wheels on the inside of the turn are stopped or even run in reverse. Many six-wheeled all-terrain vehicles (ATVs) don't even have steerable wheels and must execute all turns by running the left wheels and right wheels at different speeds. This is similar to the way in which tracked vehicles like tanks are steered. In fact it is possible to approximate the operation of treads by using closely spaced wheels as shown in Fig. 12-12.

Tracked vehicles

Bulldozers and military tanks are familiar examples of *tracked* vehicles. The various parts of a tracked drive system are labeled in Fig. 12-13. There are two fundamentally different types of tracks. The tracks used on bulldozers and tanks are made of many individual segments that are linked together to form a continuous loop. Each segment is typically a heavy metal casting with protrusions on the outer face for good traction on unpaved surfaces. The inner surface will often have a ridge or pin in its center. This pin passes through grooves in the road wheels and serves to keep the track properly positioned relative to the width of the wheels. Design and

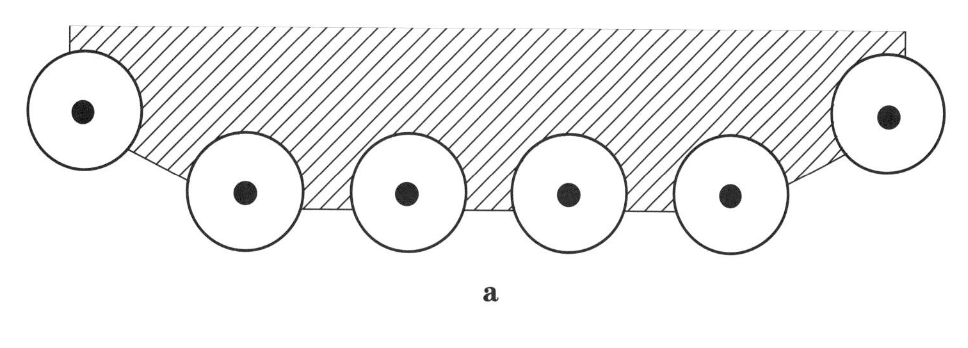

a

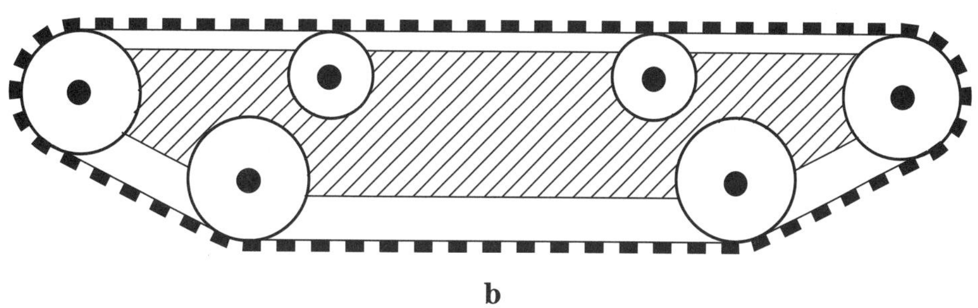

b

12-12 Closely spaced wheels in (a) approximate the operation of the track in (b).

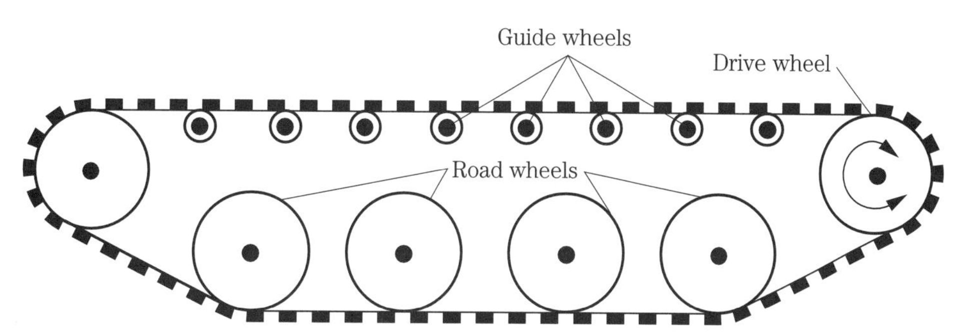

12-13 A typical track drive configuration.

construction of a linked track drive would be a major undertaking for the home experimenter. For really small robots, it may be possible to adapt a piece of bicycle chain.

The second type of track is more commonly seen on robots and toy models of tanks and bulldozers. This type of track consists of a continuous loop of some durable but flexible material—usually some type of rubber. The outer surface may have protrusions molded on or fastened on for good traction. The inner surface may be flat rubber, or there may be molded teeth that are meant to mesh with teeth on the drive sprocket. When compared to wheeled locomotion, tracked designs have some advantages and some disadvantages. The advantages include:

- Good mobility over uneven, unpaved surfaces.
- If the track is large enough, stairclimbing may be almost as easy as climbing up a similarly sloped ramp. (See Fig. 12-14.)

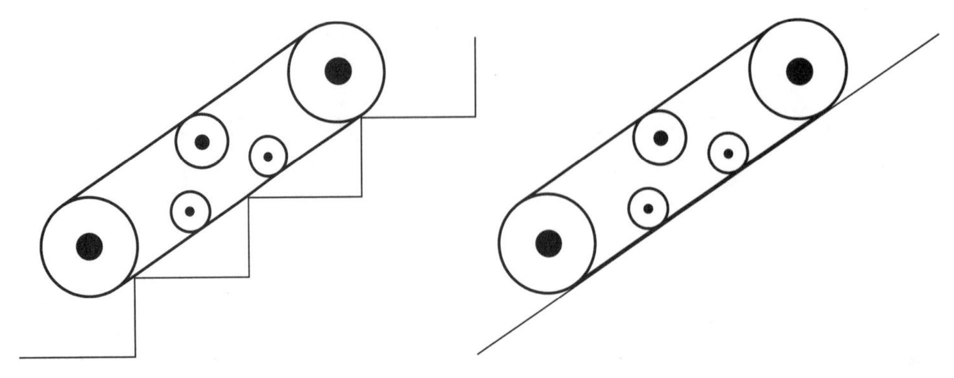

12-14 A long track can climb stairs almost as easily as it can climb a similarly sloped ramp.

- Depending on your tastes, tracked robots may have a more pleasing "high tech" appearance than do similar wheeled robots. Whether or not this matters depends on what you plan to do with the finished robot.
- A separate steering mechanism is not needed. Tracked vehicles turn by running their two tracks at different speeds and/or in different directions.

The disadvantages of tracked designs include:

- Design is more complex.
- Tracks are less forgiving of imprecise construction. If the front axle on a wheeled vehicle winds up being half an inch further forward than originally planned, it's probably not a big deal. However, if the front axle on a tracked vehicle winds up being half an inch further forward than originally planned, the track may not fit over the wheels, or if it does fit, it may bind and fail to run properly. If the front axle is further aft than originally planned, the tracks may be continually falling off.
- Linked tracks are difficult to design and fabricate.
- Continuous belt tracks may be hard to find in the exact length needed for a particular design. In most cases, it is probably best to create an initial design to determine the desired track length and then try to find something close to this size. Once the track is obtained and measured, finalize the design around actual track length.
- Although steering is "built into" track designs, the results of track steering are usually not very precise, and steering operations tend to tear up divots from soft surfaces like mud, and mark or scratch firmer surfaces like tile floors.

Track sources

Although they may be out there, I have been unable to locate a vendor that sells tracks specifically intended for small vehicle use. There are other readily available items that can be adapted for use as tracks on a home-built robot.

Toys

For smaller robots, suitable tracks can often be found in the toy store. Large metal truck toys like those made by Tonka often include rubber tracks on bulldozers and some of the other construction vehicles. "Lunar rovers" and similar space toys sometimes include tracks as well. The military vehicles set of the CONSTRUX™ building toy includes tracks and wheels like those shown in Fig. 12-15.

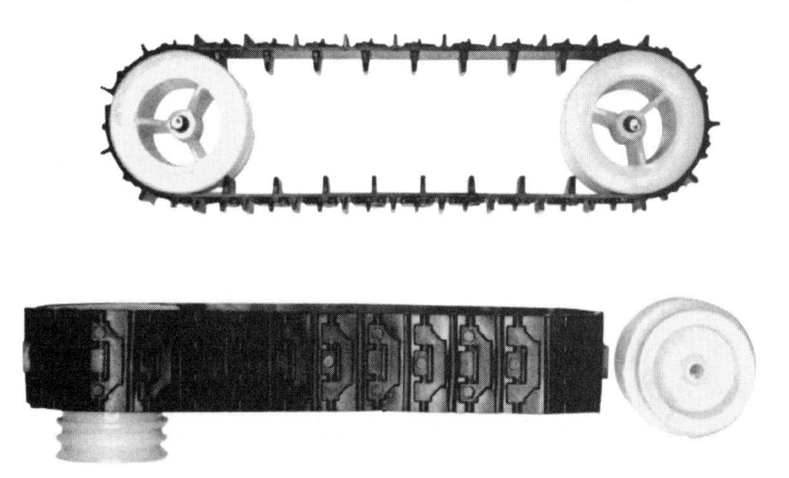

12-15 Rubber tracks and matching wheels from CONSTRUX™ building toy.

Timing belts

Flexible timing belts have often been used as tracks for small robots. Sometimes suitable belts can be scavenged from old printers, card punch machines, or other mechanical behemoths. Most devices that use timing belts tend to use sizes that are not wide enough to serve as tracks. Widths up to 1 inch are available, with the narrower widths being easier (and less expensive) to obtain. For new belts in sizes useful for tracks, figure on spending at least $25 per belt. One easy, albeit not cheap, source for timing belts is the automotive store. Although most engines use timing chains, the overhead cam engines used in Ford Escorts and similar cars use a timing belt that usually must be replaced any time the head is pulled from the engine. The belts are available from most automotive parts stores. The difficulty here may be finding affordable timing belt pulleys that mesh correctly with the belt. If purchased new from a dealer, such pulleys cost around $45 each. Most timing belts are *single-sided* with teeth only on one side of the belt. When used as a track, the teeth need to face out for traction against the ground. Because the driving pulley also needs to mesh with the tooth side, an arrangement similar to Fig. 12-16 should be used.

The large wheels at each end of the track need to have a groove to hold the track as shown in Fig. 12-17. *Double-sided* belts with teeth on both sides are available, and they can be driven from the inside as shown in Fig. 12-18. Driving the belt from the inside allows a more compact design. Furthermore, because more teeth are in mesh with inside drive, more force can be transferred to the belt without causing tooth damage. This is an advantage in heavy-duty designs. For added track power, it would be possible to add a second drive pulley near the rear of the track. In fact, it would be possible to power all of the road wheels, but the added performance would most likely not justify the added complexity and cost.

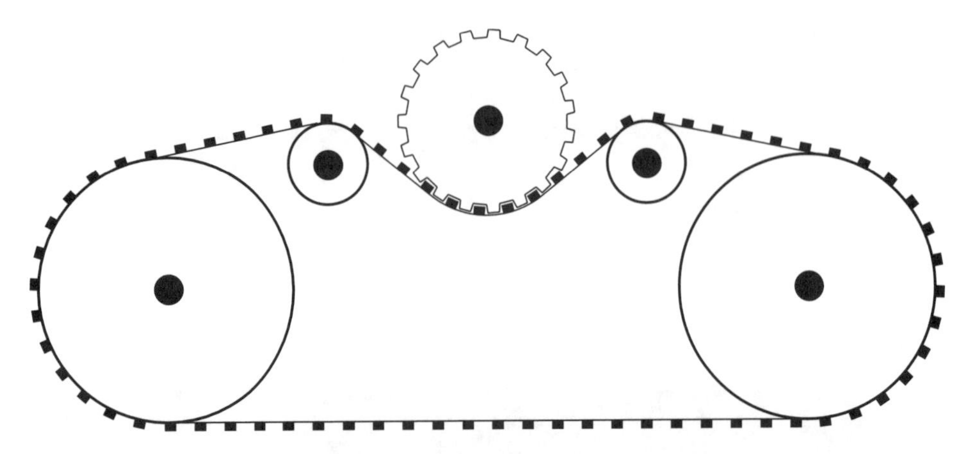

12-16 Driving and support arrangement for a track made from a single-sided timing belt.

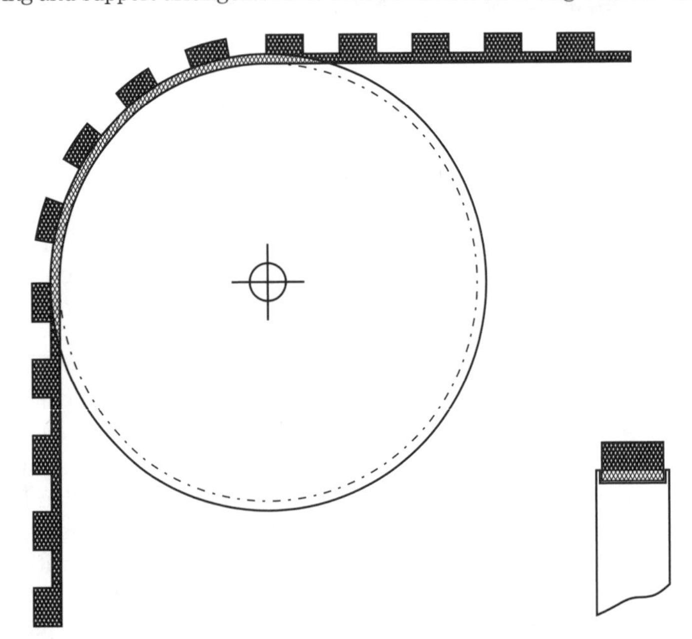

12-17 The guide wheel should have a groove just deep enough to hold the thickness of the belt measured at a point between teeth.

Bicycle tires

A serviceable belt track can be cut from the pneumatic tires used on bicycles and scooters. There are really two different approaches as sketched in Fig. 12-19. In the first approach, only the most central flat portion of a wide tire is used as the track, and the guide wheels are equipped with grooves in which the track rides. In the second approach, a slightly wider portion of the tire is taken. The portion used as the track includes some of the sidewall and the track exhibits some definite "cupping." The edges of guide wheels and road wheels must have a convex cross-section that fits snugly in the concave profile of the track's inside surface.

12-18 Driving and support arrangement for a track made from a double-sided timing belt.

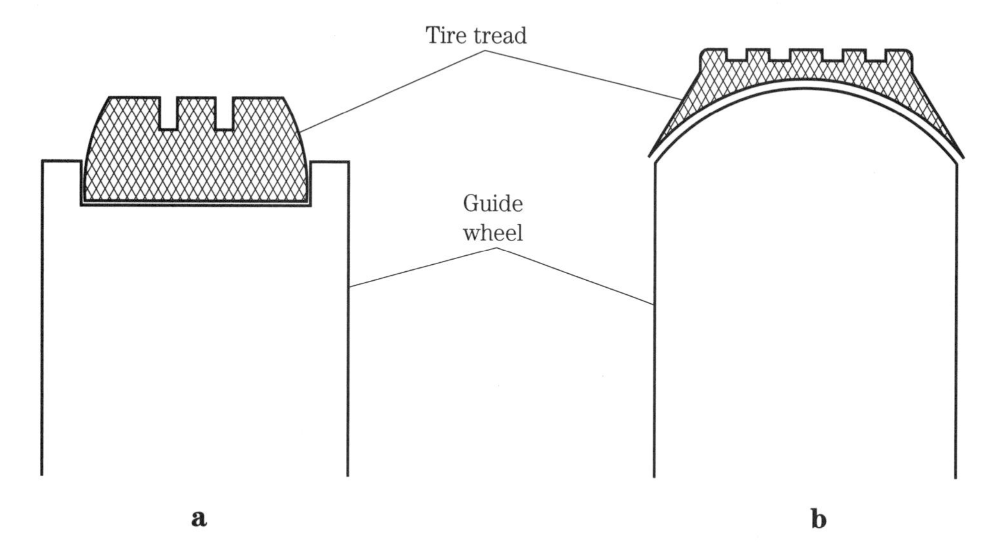

12-19 Two different ways to cut a track from a bicycle tire. In (a) a narrow flat-bottomed strip rides in a flat groove around the edge of the guide wheel. In (b) a concave-bottomed strip rides over the convex edge of the guide wheel.

Depending on the amount of sidewall left on the track, the guide wheels may or may not need to have flanges in order to keep the track on the wheels. The more concave the track is, the less likely it is that flanges will be needed. If too much sidewall is retained, it may be necessary to cut notches as shown in Fig. 12-20 so that the track can wrap snugly around the guide wheels without buckling. The white and brightly colored tires sold for small-wheeled bikes are made from a rubber that seems to be more flexible and "tackier" than the usual black rubber. These characteristics are definite advantages when the tires are used as tracks. They wrap around the guide wheels more easily, and they grip better on tile floors. Besides, bright orange or poison green tracks will have more of a cool "techy" look than tracks cut from plain old black tires.

Driving over obstacles

The raised front wheel and raised front of the track in Fig. 12-12 aid in driving up and over small obstacles. As shown in Fig. 12-21, an obstacle that is too tall will

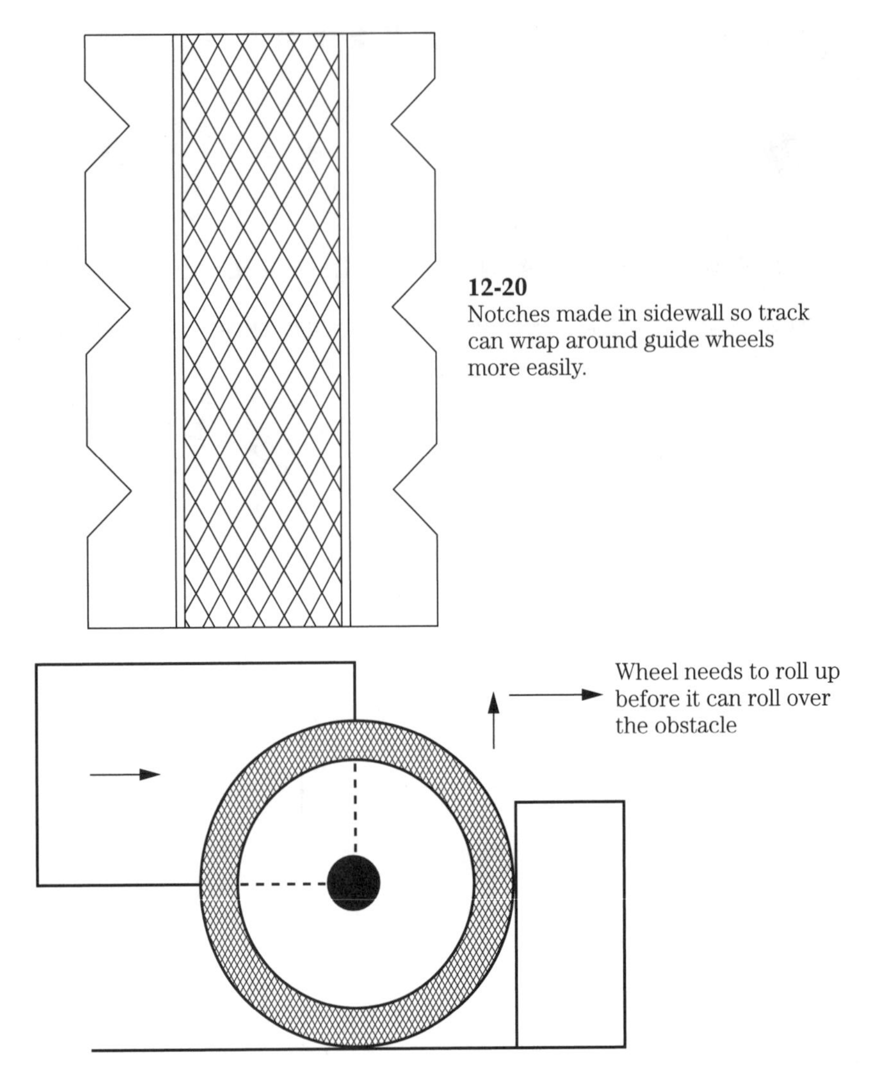

12-20 Notches made in sidewall so track can wrap around guide wheels more easily.

12-21 When the front wheels encounter a tall obstacle, the vehicle stops unless the front wheels can climb vertically up the obstacle.

require the front wheels to climb **vertically** up the face of the obstacle before any forward motion **over** the obstacle can begin. Such an obstacle will effectively block most vehicles. For obstacles lower than the centers of the front wheels, the corner of the obstacle will hit the wheel somewhere in the lower front quadrant as shown in Fig. 12-22. The wheel will roll up and over the obstacle in one combined motion. Raising the front wheels as in Fig. 12-12 will simply allow the vehicle to roll up and over larger obstacles.

A number of special wheel configurations have been used on vehicles and robots to provide improved performance against obstacles. Because most industrial robots operate in benign environments (smooth floors, ramps, or elevators instead of stairs, etc.), examples of mobile robots designed to handle obstacles are most easily found in movies or toy stores.

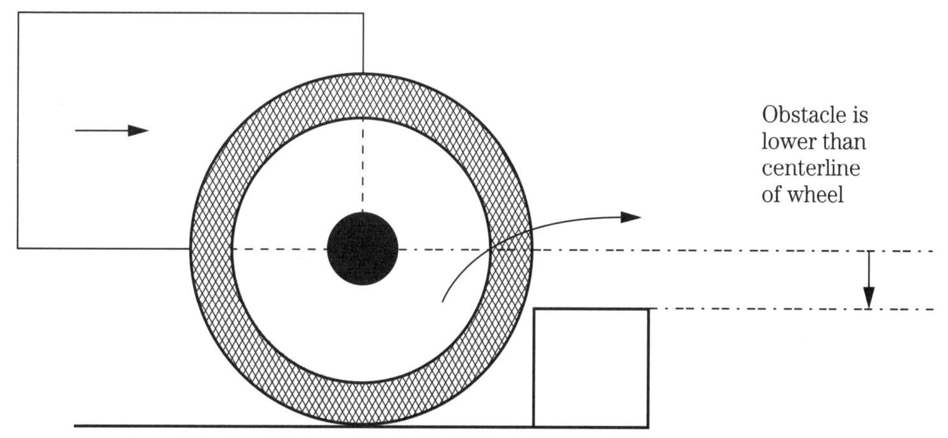

12-22 When the front wheels encounter an obstacle lower than the centers of the wheels, the wheel will roll up and over the obstacle in one combined motion.

Tri-star wheels

In the movie *Damnation Alley*, survivors of a nuclear holocaust travel cross country in a vehicle that uses *tri-star* wheels like the ones shown in Fig. 12-23. All three wheels are driven all the time, but only two wheels contact the ground at any one time. The star is free to rotate around its center pivot, but rotation of the star only occurs in response to forces caused by contact with obstacles; the rotation is not powered or commanded from the vehicle. According to one source, this tri-star wheel arrangement is based on a design patented by Lockheed Aircraft for an all-terrain vehicle.

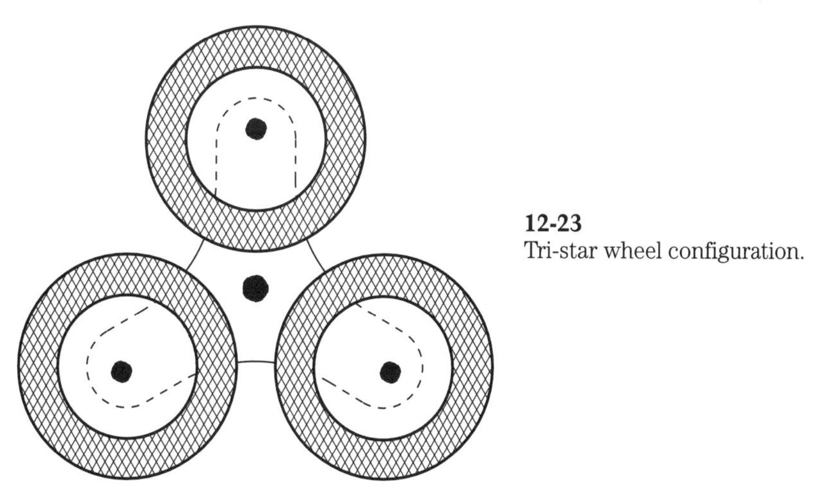

12-23
Tri-star wheel configuration.

Tiltable tracks

The robots featured in the movie *Short Circuit* each ride on two track drives plus a stabilizer wheel that is trailed to the rear. Each of the track drives is configured as shown in Fig. 12-24. The centers of the three wheels lie at the vertices of a triangle. As shown in Fig. 12-25, the tracks are connected to the rest of the body via

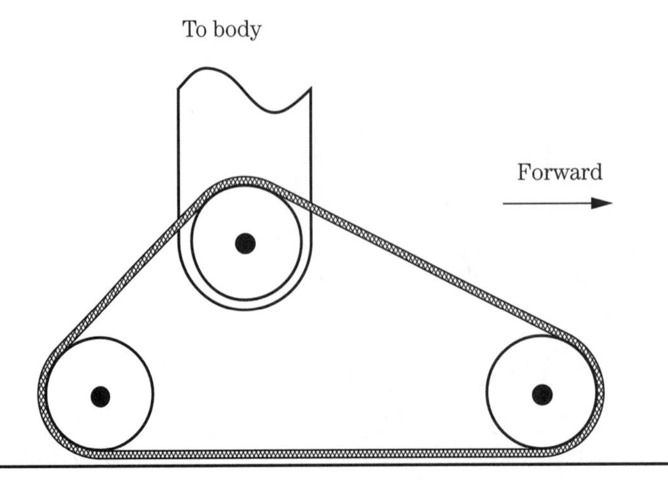

12-24 The triangular track used on the robots in the movie "Short Circuit." The track is shown in the normal "flat-footed" position used for normal running.

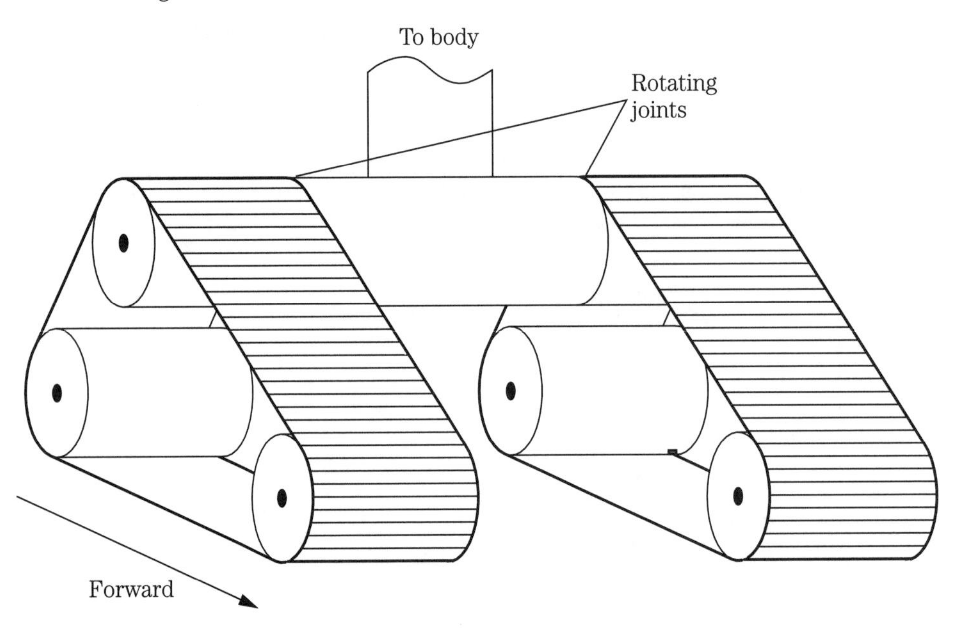

12-25 Two triangular track drives shown pivot-mounted to the body stem of the robot.

a rotating joint that pivots around the centerline of the top track wheels. The tracks can be pivoted from the position shown in Fig. 12-24 to the position shown in Fig. 12-26. The "flat-footed" position of Fig. 12-24 is used for "normal" running. The "tip-toe" position of Fig. 12-26 is used for maneuverability in close quarters. Although I failed to see it during repeated viewings of the movie, I imagine it would be useful if the tracks could also pivot to the position shown in Fig. 12-27. This position should allow the leading edge of the track to get up over the edge of a normal stairstep.

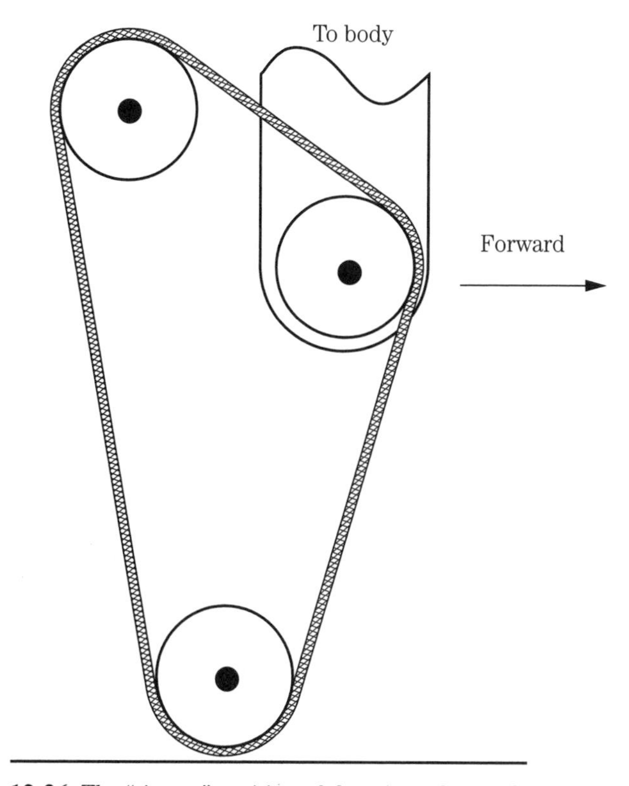

12-26 The "tip-toe" position of the triangular track.

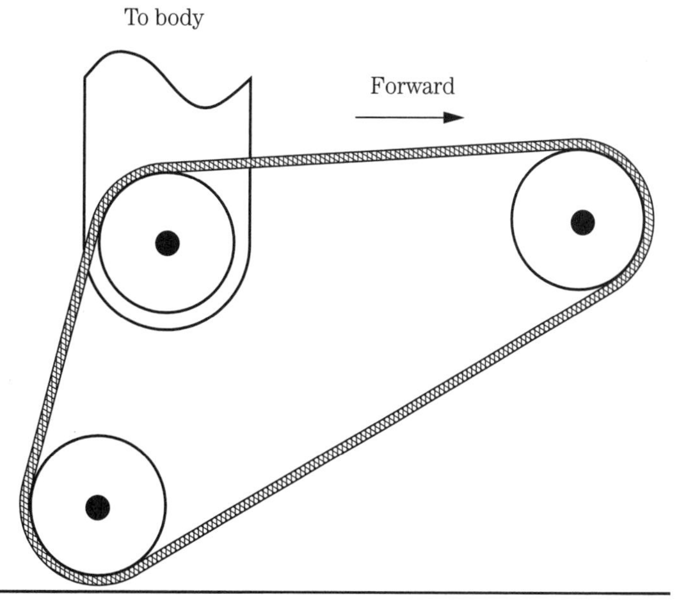

12-27 A "heel-down/toes-up" position of a triangular track that could be useful for stair climbing.

Overdrivers™

Figure 12-28 shows one of the four different vehicles called **Overdrivers™**, which are made by Tiger Electronics Inc. The particular model shown is called **The Savage**. The other three models are **Road Menace**, **The Clasher**, and **Phantom**. The rear wheels are ordinary wheels about 3 inches in diameter with the outer 0.3 inch comprising a band of deeply toothed rubber. These wheels are rigidly attached to their axles which are driven by the geared-down output from the motor. The front "wheels" are ovals that measure about 3.2 inches by 5.5 inches. The outer 0.3 inch is a band of deeply toothed rubber similar to the rear wheels. Inside of the oval there is an oval-shaped internal gear which measures about 2.2 inches by 4.5 inches. Attached to each end of the front axle there is a gear with a diameter of about 1.9 inches. These gears mesh with the oval internal gear in the front wheels. Let's assume that the vehicle starts out in the position shown in Fig. 12-28.

12-28 Photograph of an Overdriver™ toy car showing its unique wheel design.

As the axle turns, the gear rolls forward inside the oval wheel until it reaches the forward circular end of the internal gear. As the gear continues to roll forward, it causes the rear of the oval wheel to lift off the ground as shown in Fig. 12-29. Continued forward motion will cause the rear end of the oval to swing over top of the axle and become the forward edge, bringing the vehicle to the position shown in Fig. 12-30. If there are any small obstacles in front of the vehicle, the front edge of the wheel will come down on top of the obstacle, thus allowing the vehicle to pull itself up onto the obstacle rather than running into the side of the obstacle and being stopped. The internal details of the vehicle are shown in Fig. 12-31. It seems like a fairly straightforward matter to build a small robot body that incorporates this chassis.

The Overdrivers™ are too small to climb normal size stairs, but the same type of oval wheel could be constructed in a larger version. The internal oval gear appears to be the most difficult part to fabricate in the home shop. An alternative design to accomplish the same sort of operation could substitute a bicycle sprocket for the cir-

12-29
Overdriver™ with rear of oval wheel lifted off the ground.

12-30 Overdriver™ with forward edge of wheel poised to come down on top of an obstacle.

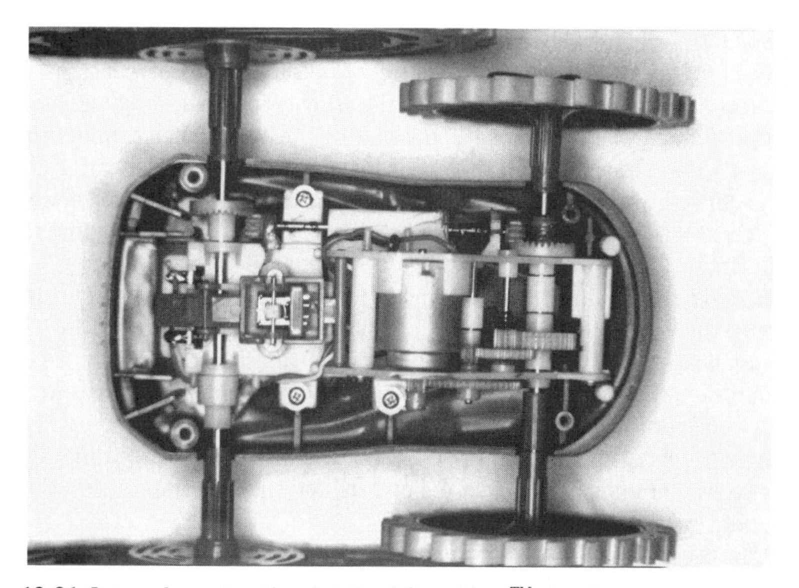

12-31 Internal construction details of Overdriver™ chassis.

cular gear, and use a loop of bicycle chain fastened inside an oval wheel frame in place of the internal gear. In settling on the sizes for such a wheel, it is important to make sure that several critical relationships between various dimensions are maintained. Refer to Fig. 12-32.

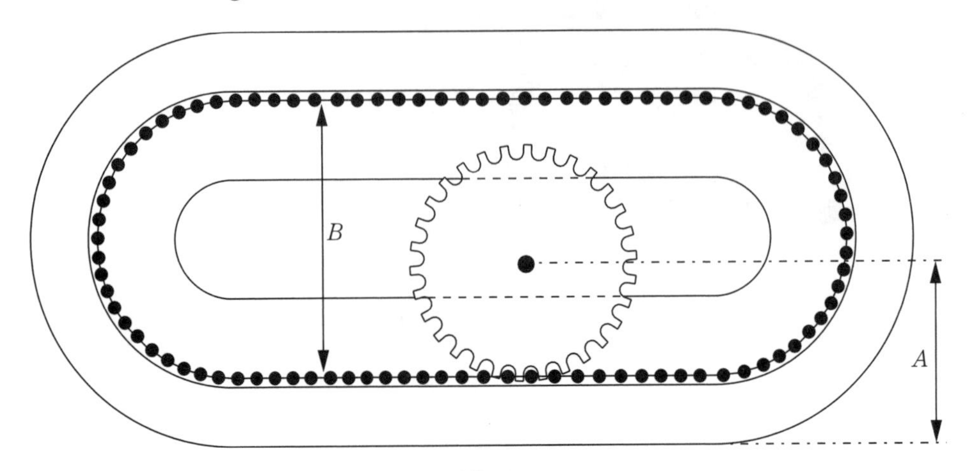

12-32 Critical dimensions for an Overdriver™ wheel.

The distance **A** needs to be large enough to provide adequate ground clearance for the body of the vehicle. The distance **B** needs to be large enough to allow adequate clearance for the top of the sprocket to clear the top length of chain as it rolls forward in the bottom length of chain. The size of the hole in the wheel needs to be large enough that the axle of the sprocket clears the edge of the hole as the sprocket meshes with the chain at every point around the oval. The hole also needs to be small enough that the top of the gear cannot pass through the hole when the bottom of the gear is meshed with the bottom length of chain. The circumference of the inner edge against which the chain will be placed must be such that a whole number of links will fit exactly along this oval. A good procedure for obtaining the correct dimensions is:

1. Select approximate dimensions for the wheel's internal track and assemble a loop of chain that is as close as possible to the selected circumference.
2. Lay the loop of chain on a sheet of heavy paper and carefully arrange it into the correct oval shape. Use a compass to draw the semicircular ends of the internal track and carefully lay the **outer** edge of the chain against the drawn outlines. Several attempts at tweaking the dimensions may be needed to come up with an outline in which the chain loop is an exact fit.
3. Once the correct outline for the internal track is determined, transfer this outline to a piece of 1"×8" or 1"×12" fir or clear pine board. Because of the router operations to follow, it would not be a good idea to substitute plywood for the 1" board stock.
4. Draw the external wheel outline around the internal track outline. The semicircular ends of the external outline and the internal track outline should share common centers.
5. Draw the outline for the through hole, making sure to satisfy the relative sizing requirements discussed above. At this point, the board should have on it outlines for three "concentric" ovals. The largest oval is the outer edge

of the wheel, the smallest is the outline of the through hole, and the intermediate oval is the outline of the outer edge of the internal track that will hold the loop of bicycle chain.

6. Cut out the through hole using a drill and saber saw or coping saw. It will be easier to hold down the wood for the routing operations to follow if the wheel is not cut out until later. (There must be enough external scrap to clamp down the wood without having the clamps get in the way of the router's baseplate.)
7. Using a router, reduce the thickness of all the area inside the middle oval to produce a "step." The loop of bicycle chain will be placed against the oval edge of this step as shown in Fig. 12-33.

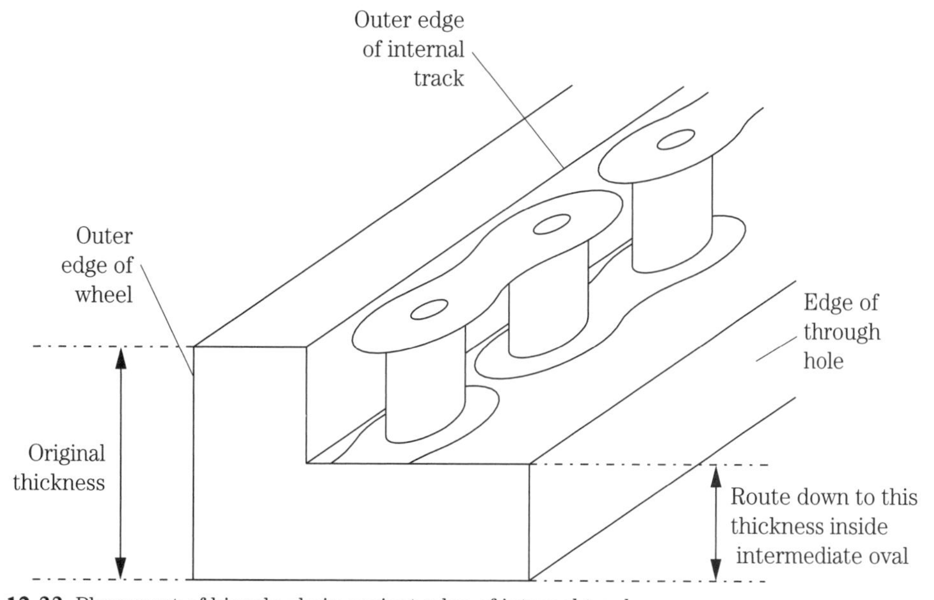

12-33 Placement of bicycle chain against edge of internal track.

8. Cut out the external outline of the wheel. Sand and paint as desired. It will be easier to paint the wood before the chain is attached.
9. Cut the tread strip from a bicycle tire. Attach this strip around the outer edge of the oval wheel using glue and nails. (The ribbed nails used for panelling will not pull out as easily as smooth common nails.)
10. Attach the chain against the edge of the oval step. The best way to attach the chain is somewhat tedious. Every third link or so, drill out the pin used to join adjacent links. Replace this pin with a ribbed nail that is driven into the wood as shown in Fig. 12-34. This operation will be easier if the entire chain is separated into three-link sections before any of the chain is attached to the wheel.

If a router is not available, the wheel can be assembled by fastening together two different layers. One layer has an internal through hole the same size as the finished wheel's internal through hole. The second layer has an internal through hole the same size as the outer edge of the internal track.

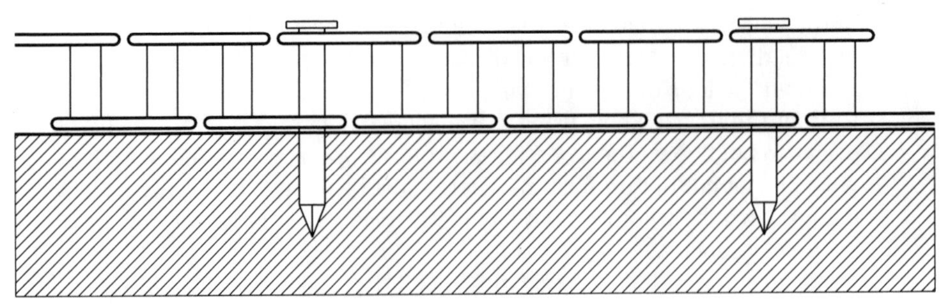

12-34 Attachment method for bicycle chain.

13
CHAPTER
Arms, legs, and hands

This chapter shows how the mechanical devices presented in previous chapters can be combined to form various parts of robot anatomy.

Wrist joints

It has become almost standard to use a gear differential to control the motion of a robot's wrist. Figure 13-1 shows a simple demonstration joint that can be assembled using nothing but parts from a Gears-In-Motion set. The salient features of this joint are diagrammed in Fig. 13-2. If both drive pulleys are rotated at the same speed in the same direction (say tops towards the gripper), the left gear will tend to make the gripper rotate counterclockwise as viewed from the arm, and the right gear will tend to make the gripper rotate clockwise as viewed from the arm.

The net result is that the gripper will not rotate, but because both drive gears are pushing down on the middle gear, the middle gear, gripper bracket, and gripper will all pivot down around the two drive-gear axles. This motion is called *pitch*, and the centerlines of the two drive axles define the *pitch axis*. If the two drive pulleys are rotated at the same speed but in opposite directions, the middle gear will turn causing axle **a** to rotate the gripper. This rotation is called *roll*, and the centerline of the middle pulley's axle defines the *roll axis*. If the two gears are rotated at different speeds, the gripper will move in some combination of pitch and roll. The precise combination depends on the relative speeds of the drive pulleys. The roll axis and pitch axis are perpendicular to each other, and a third axis can be defined that is perpendicular to both of them.

Motion centered about this third axis would be called *yaw*. The normal human wrist is capable of yaw and pitch motions, but virtually all of the roll motion of the hand is generated in the forearm rather than in the wrist itself. (If you don't believe this, try grabbing your forearm just above the wrist. While holding your forearm still, try to roll your hand. Even though yaw and pitch motions are easy, roll will be virtually impossible.) For many robot grippers, pitch and yaw are produced by a wrist differential, and the equivalent of yaw is produced with arm movements or torso

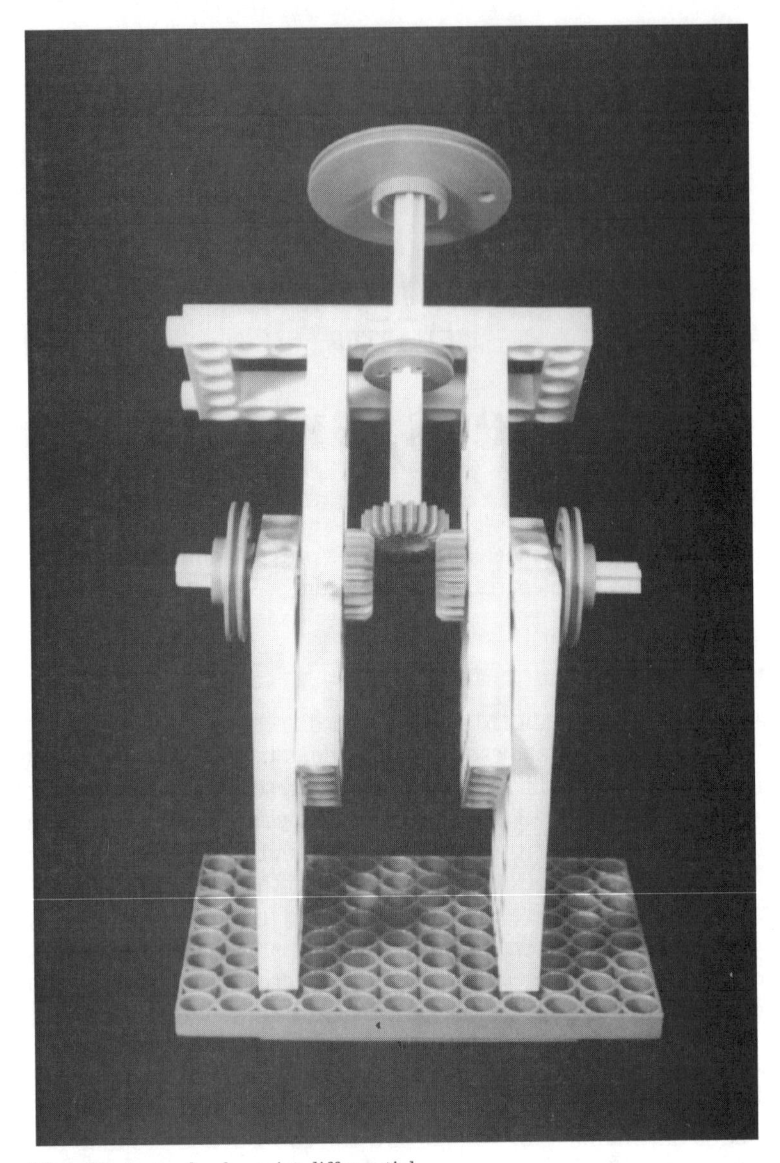

13-1 Photograph of a wrist differential.

rotations. Roll is unique, but yaw and pitch could be considered interchangeable; if we rotate the entire wrist assembly by 90 degrees around the roll axis, so that the pulleys are on top and bottom rather than on the sides, then yaw becomes pitch and pitch becomes yaw. This rotated design is sometimes used because it is often easier to design an elbow that supports equivalent pitch motion than one that supports equivalent yaw motion. The pulleys that drive the wrist are driven by belts that can extend well up inside the arm.

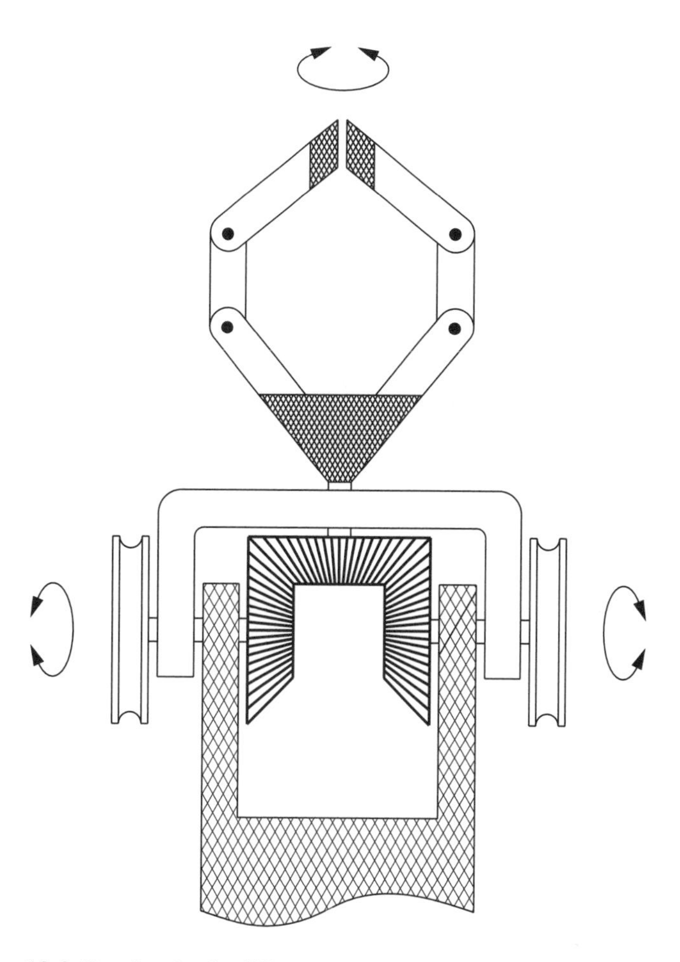

13-2 Details of wrist differential.

Walking

Legged robots can be grouped into two classes: *statically stable* walkers and *dynamically stable* walkers. Statically stable walkers can stop at **any** point in the walking process and remain stable for an indefinite period of time. Dynamically stable walkers depend on momentum and the laws of motion to smoothly "carry them through" periods of potential instability while moving from a stable initial position to a stable final position. If a dynamically stable walker were to stop suddenly in mid-stride, it might fall flat on its face. We humans exhibit both statically stable and dynamically stable modes of locomotion. Running is almost totally dependent on dynamic stability; at some instants, neither of a runner's feet is in contact with the ground. The slow careful steps of a senior citizen are an attempt to make the walking process statically stable. Normal walking is a mixture of both kinds of stability. Most of us could stop in mid-stride and not fall over, at least not immediately. After a few seconds of standing on one foot, we start to wobble and tense the appropriate leg muscles in an attempt to achieve static stability.

For static stability, the center of gravity must be directly above a point that lies inside the region defined by the points of ground contact. The points of contact must define a region; it is not possible to achieve static stability with one or two points of ground contact, or even with a larger number of points if these points all lie on the same line. It is possible for humans to stand on one leg because even when only one foot is on the ground, we still have more than one **point** of contact. Figure 13-3 shows the footprint of a typical human foot. The darkest areas show where the naked foot exerts the most pressure on the ground. For static stability, the body's center of gravity must be directly above the black or gray shaded area. This is not a very large target to hit, and the posture needed to correctly position the center of gravity can quickly become uncomfortable.

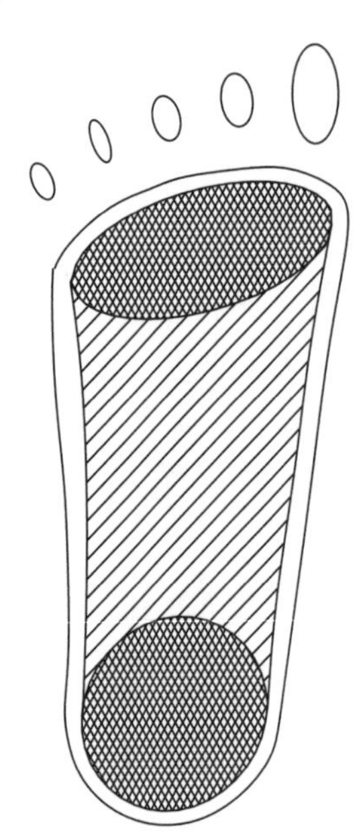

13-3
Ground-contact areas of a typical human foot.

Statically stable and dynamically stable are the two classifications of "true" robot walking in which legs are lifted clear of the ground and placed down in new locations. There is a third category of walking employed in some toy robots and other types of walking toys in which the legs are never lifted clear of the ground. In this pseudo-walking, the feet are "shuffled" back and forth across the ground without being lifted. Usually some sort of ratcheting device is built into the bottom of each foot so that when the foot moves forward it slides over the ground surface without imparting any motion to the robot body. When the foot moves backward (relative to the body), the ratcheting device engages, causing friction to hold the foot still relative to the ground thereby causing the body to move forward. The ratcheting devices are usually wheels that roll easily in the forward direction but lock up in the reverse direction. Sometimes, in the **really** cheap toys, ridged rubber is mounted on the bottom of each foot in such a way that when the foot moves forward, the ridges collapse and fold flat as shown in Fig. 13-4, allowing the foot to slide easily over the surface. When the foot moves backward, the ridges open up and "bite into" the surface, causing a high-friction interface between the foot and the surface. In the discussions that follow, the emphasis will be on each configuration's suitability for "true" walking rather than this type of pseudo-walking.

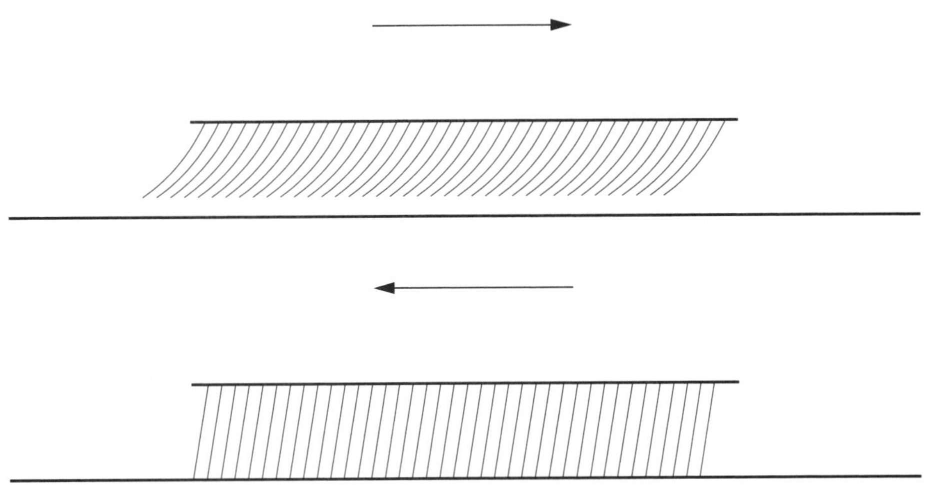

13-4 Rubber ridges used on some walking toys to provide "one-way" friction.

One leg

We could design a one-legged robot with a footprint large enough to guarantee that it covered the point directly below the robot's center of gravity, and would therefore be statically stable. However, such a robot wouldn't walk and most people would call the leg a pedestal. Humans can hop on one foot, and so can some robots. Researchers at Carnegie-Mellon University have developed a **dynamically stable** one-legged robot (with a small foot) that hops from place to place.

Two legs

The ultimate humanoid robot would walk on two legs (be *bipedal*) and be capable of negotiating different types of terrain to include stairs. Unfortunately, true bipedal dynamically stable locomotion over diversified terrain is among the most difficult of mechanical behaviors to design into a robot. Most two-legged robots that have been built rely on some degree of static stability provided by large feet and carefully controlling the center of gravity's position.

Three legs

In geometry, it takes three points to define a plane. Therefore, three is the minimum number of legs that can provide static stability. As long as the center of gravity's *projection* onto the ground lies within the triangle defined by the points-of-contact for the three legs, the robot will be statically stable. (The center of gravity will actually lie somewhere inside the robot's body, but the discussions of stability presented here only "care" about the location of the surface point vertically below the CG. This is the CG's *projection* onto the ground surface. In subsequent discussions, we will loosely refer to this projection as the CG, realizing that the actual CG is some vertical distance away inside the robot's body.) However, if the robot is to move anywhere, the legs must move and there will be times during which only two legs are in contact with the ground. During these times the robot may be potentially unstable.

Four legs

Four is the minimum number of legs that can provide static stability **even while one of the legs is moving to a new position.** The mathematics and physics involved in maintaining dynamic stability is beyond the scope of this book, so let's see what it would take to ensure uninterrupted static stability in a four-legged walking robot. (Bear in mind that any scheme that delivers uninterrupted static stability will not be able to reproduce the natural gait of a four-legged animal and will always look somewhat "stiff" and "mechanical.") Let's assume that the legs are attached to the robot body at four points that define a rectangle. Each leg is capable of contacting the ground at a point that could be either forward or aft of the leg's attachment to the robot. Each contact point is also some fixed distance outboard from the side of the robot's body.

Figure 13-5 shows a top view of a four-legged robot, with the range of possible contact points shown for each of the legs. For the sake of a concrete example, let's assume that the robot's center of gravity lies at the geometric center of the body as shown in the figure. Bear in mind however, that unless special balancing measures are employed, the center of gravity will most likely not lie exactly at the geometric center. A particularly massive component such as a car battery could shift the center of gravity by a significant amount. In fact, for any robot with moveable appendages like arms, legs, or camera booms, the center of gravity of the unit will move around depending on the positioning of these external appendages.

Although four legs have the potential to provide static stability, realizing this potential is not particularly easy. Let's assume that the robot is standing at "attention" with each leg being vertical contacting the ground directly below the leg's pivot point. We would like to begin walking by moving the left front leg out to its most forward position. For static stability to be maintained throughout the duration of this move, the robot's center of gravity must lie inside the triangle formed by contact points **LR**, **RR**, and **RF** as shown in Fig. 13-6.

The center of gravity shown lies directly on the edge of this triangle. At best, the robot would be marginally stable. However, even if accurately placed for the "attention" posture, the center of gravity shown will move forward as the left front leg moves forward. This will cause the robot to lean toward the left front. What can we do? Let's intentionally shift the internal weight distribution so that the at-rest CG is moved to the rear as shown in Fig. 13-7. Now the CG will stay inside the (**LR**, **RR**, **RF**) triangle as the left front leg moves to the position shown in Fig. 13-8. Assume that we wish to duplicate as much as possible a human's crawling motion in which the left hand and right leg move together and the right hand and left leg move together. (Simultaneously moving two of four appendages involves dynamic stability, so we will have to settle for moving the left forward leg and right rear leg sequentially rather than simultaneously.)

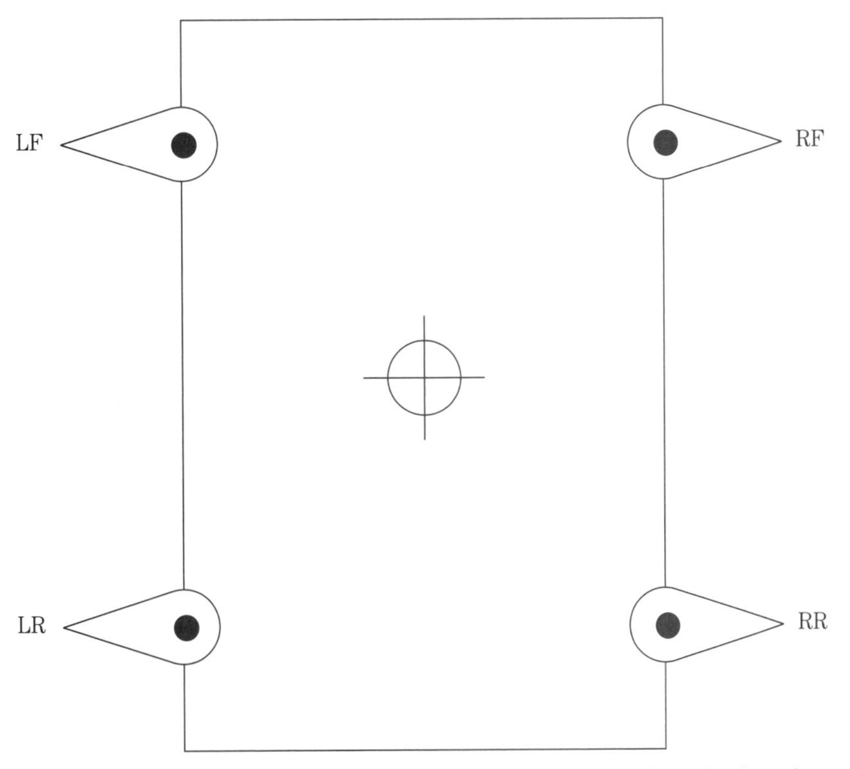

13-5 Top-view schematic depicting leg positions and center of gravity for a four-legged robot.

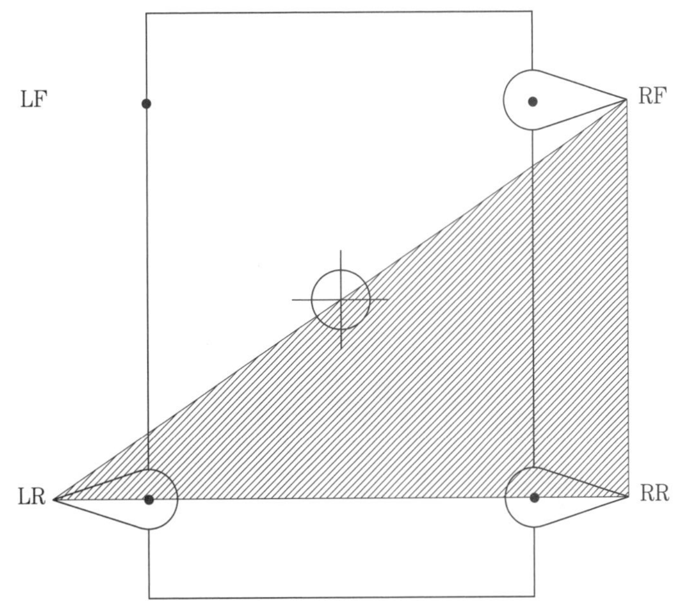

13-6 Schematic depicting marginally stable condition where CG lies on the edge of the triangle defined by the three support points.

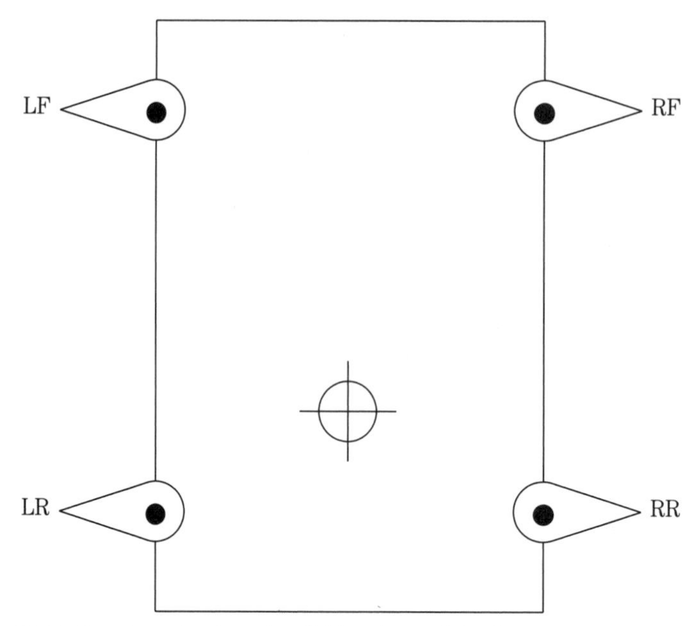

13-7 Top view of robot after CG has been shifted towards the rear.

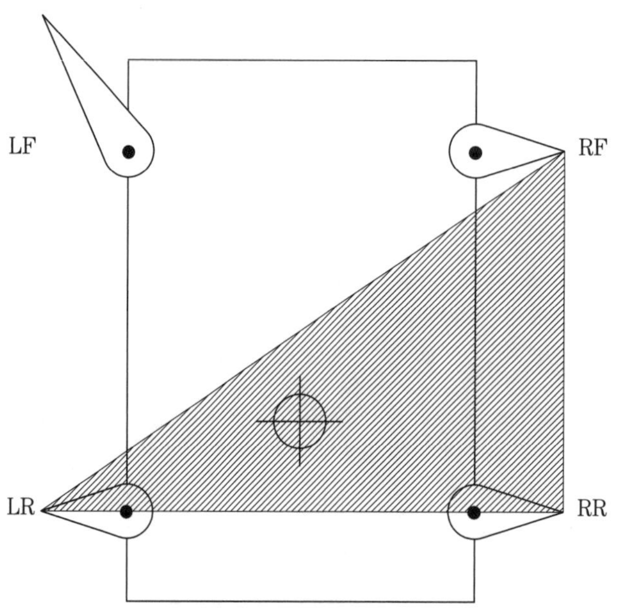

13-8 Top view showing stable location of CG and new position for left front leg.

To evaluate stability for moving the right rear leg to its most forward position, we must examine the CG's relationship to the triangle formed by points **LF**, **LR**, and **RF** as shown in Fig. 13-9. The CG is outside the triangle so the robot will be unstable if the right rear leg is lifted off the ground. Let's try a different strategy by moving the right front leg forward instead of the right rear leg.

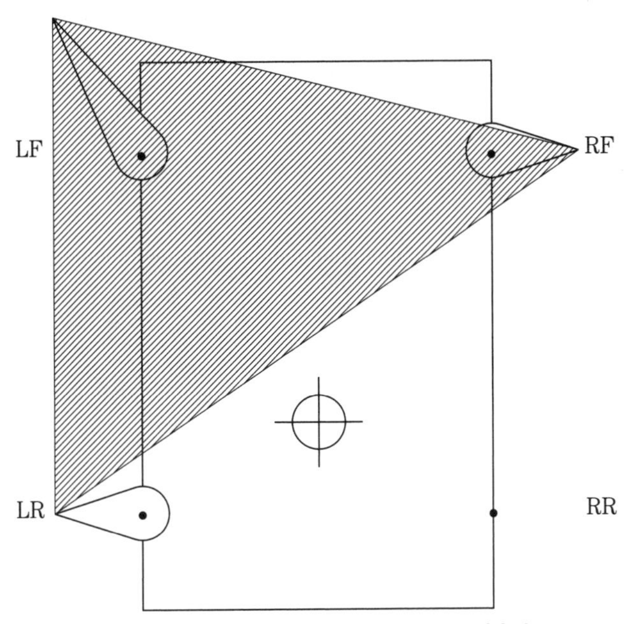

13-9 Top view showing unstable location of CG for movement of the right rear leg.

The CG falls inside the (**LF**, **LR**, **RR**) triangle, so the stability will be maintained while the right front leg moves forward to the position shown in Fig. 13-10. Because the CG is towards the rear of the robot, stability will be lost if either of the rear legs is lifted off the ground. What we can do is shift the body forward while leaving all four feet firmly planted on the ground. This will result in the configuration shown in Fig. 13-11. Now we need to move the rear legs forward one at a time to the starting position of Fig. 13-7. The whole sequence of movements can now be repeated as many times as necessary to move forward by the desired amount.

How does a four-legged robot turn? With only the simple forward-aft motions discussed so far, only very crude "skid turns" can be accomplished. These turns enormously complicate navigation due to their inaccuracy, and if the robot is heavy enough they can tear up divots from the front lawn or living room carpet. Precise clean turns require that the legs have the capability for some side-to-side as well as front-to-back movement.

Six legs

A very popular approach for designing robots that walk appears to have been inspired by the six-legged walking behavior of insects. Having six legs makes it

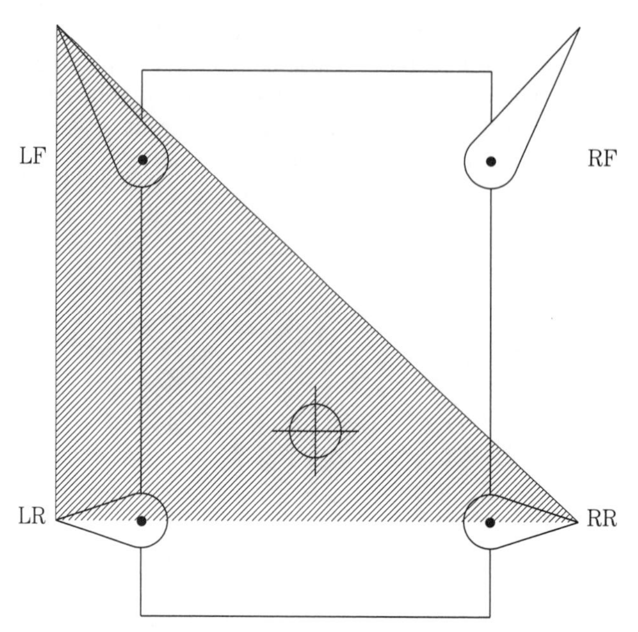

13-10 Top view showing stable location of CG and new position for right front leg.

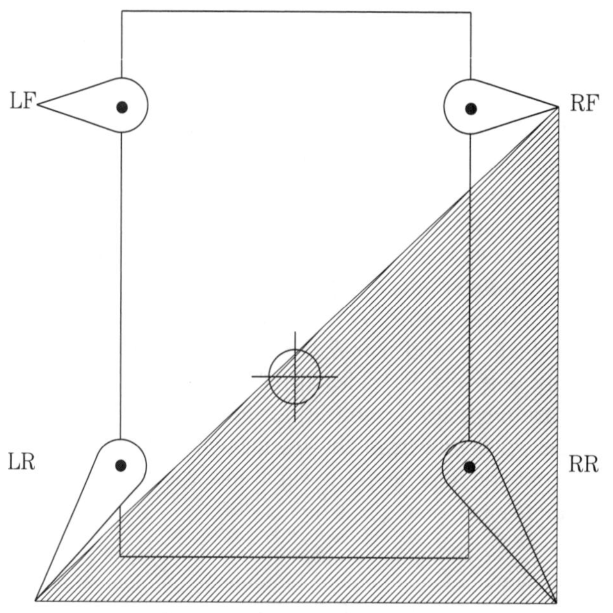

13-11 Top view showing leg positions after body is shifted forward.

easy to maintain three points of contact with the ground at all times. Three legs support the robot while the other three legs are moving to new positions. By supporting the robot with the center leg from one side with the front and rear legs from the other side, any reasonably located CG is virtually guaranteed to lie inside the triangle defined by the three points of ground contact. Once the three moving legs are planted in their new ground positions they support the robot while the other three legs advance to their new positions. It is actually fairly simple to implement six-legged walking for reasonably smooth surfaces. The key to this simplicity lies in a six-legged robot's ability to use a simple kneeless leg mechanism that provides continuous leg motion having a constant up-down duty cycle and a constant forward-backward duty cycle. There is a visual similarity between this type of six-legged gait and the gait of some four-legged pseudo-walkers, but in the six-legged case the legs are being lifted clear of the ground before being moved forward.

Kneeless leg mechanism

Consider the mechanism shown in Fig. 13-12. Joint **A** is a prismatic joint (see chapter 8) and joints **B** and **C** are revolute joints. As crank **D** rotates in a complete circle around joint **C**, the pin of joint **A**, which is fixed to link **E**, will slide back and forth in the slot cut into link **F**. The bottom end of link **E** will trace out an oval path as sketched in the figure. The vertical distance d_V between the lowest and highest points on this path equals twice the distance between pivot **C** and pivot **B**:

$$d_V = 2 \times d_{BC}$$

The horizontal distance d_H between the leftmost and rightmost points on this path is given by:

$$d_H = \frac{2 \times d_{BC} \times d_{AX}}{d_{AB}}$$

where

d_{BC} = distance between pivot **B** and pivot **C**
d_{AX} = distance between pivot **A** and the end of link **E**
d_{AB} = distance between pivot **A** and pivot **B**

The mechanism shown in Fig. 13-12 can be used as leg for a robot with the following provisions:

- Link **F** is rigidly attached to the robot body.
- The pin of joint **C** is rigidly attached to link **D** and allowed to pivot in link **F**. The pin is driven by a motor so that link **D** continuously rotates around the joint.
- The bottom of link **E** is fitted with a rubber pad or other suitable "shoe" for making a good high-friction contact with the ground.
- As the shoe traces out its oval path there will be a point in its downward motion that the shoe comes in contact with the ground.

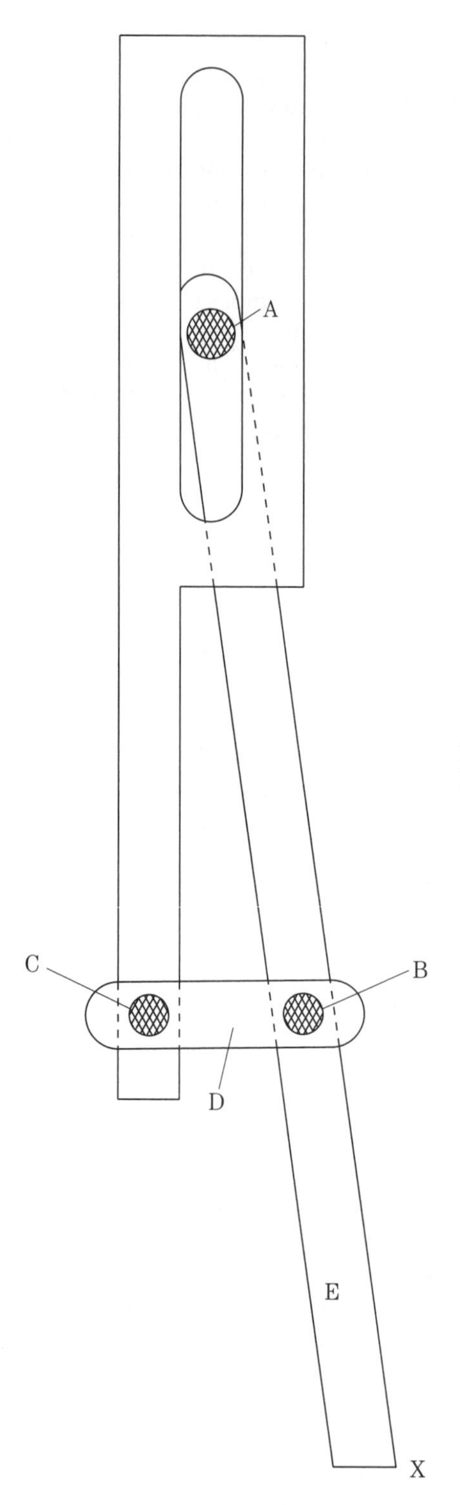

13-12
Linkage used to implement a kneeless leg.

Stair climbing

Some roboticists consider the ability to walk up normal stairs as the ultimate test of robot mobility. As discussed in chapter 12, treads or specialized compound wheels can be used to climb stairs, but in this section we are concerned only with ways that a robot **with legs** can walk up or down a flight of stairs. As shown in Fig. 13-13, a typical staircase will have steps with 7.5-inch risers and 11-inch treads.

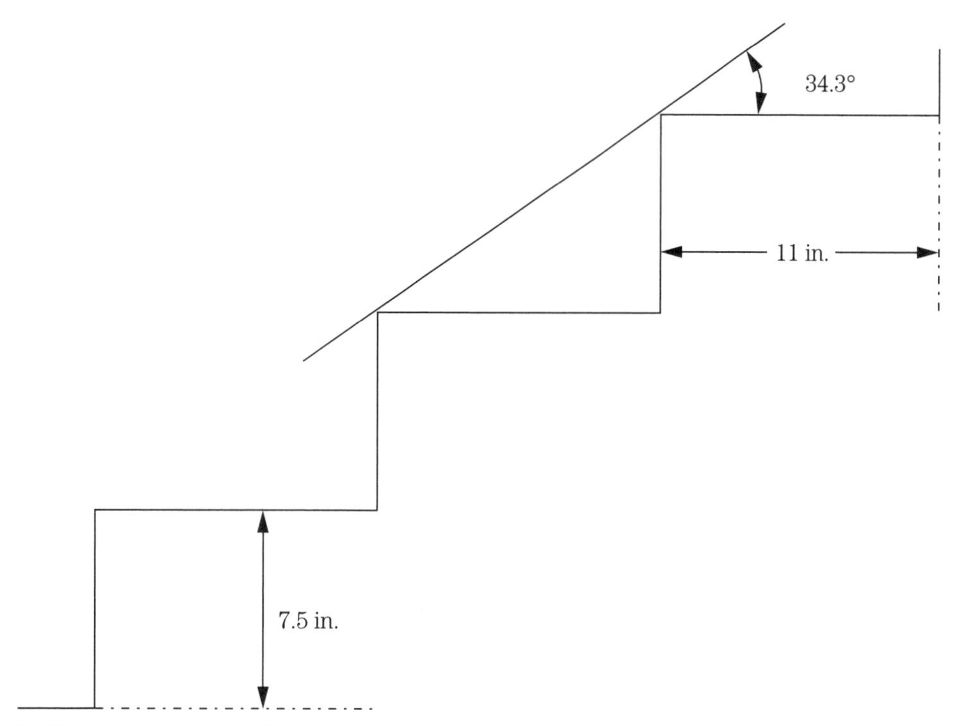

13-13 Dimensions of typical stairs.

If a board is laid across the edges of the steps to form a ramp, the ramp would be inclined 34.3 degrees from horizontal. In exploring ways to design stair-climbing robots, we must first decide what the requirements really are. Is it okay for the base of the robot to tilt severely as shown in Fig. 13-14, or must the base remain nearly horizontal as shown in Fig. 13-15? (Only the right legs of each robot are shown. It is assumed that each of the left legs is parallel to, and thus hidden behind, the corresponding right leg.)

Some robot designs, such as the one depicted in the figures, may appear to be usable in either situation, requiring only software changes to the program for controlling the sequence of individual leg motions. Appearances can be deceiving. Both figures depict a robot **standing** on stairs; neither figure adequately conveys the complexity of actually **walking** up the stairs.

In sketching the two schemes, I intentionally imposed the additional constraint that the lower portion of each leg be perpendicular to the surface of the step so that

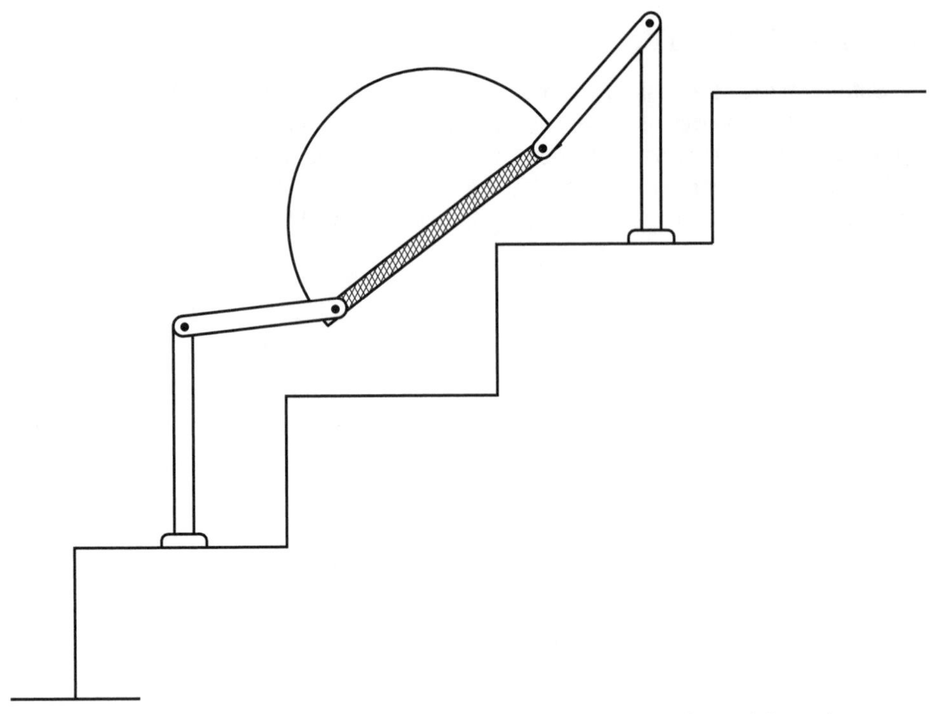

13-14 Four-legged robot on stairs with body tilted to approximate slope of the stairs.

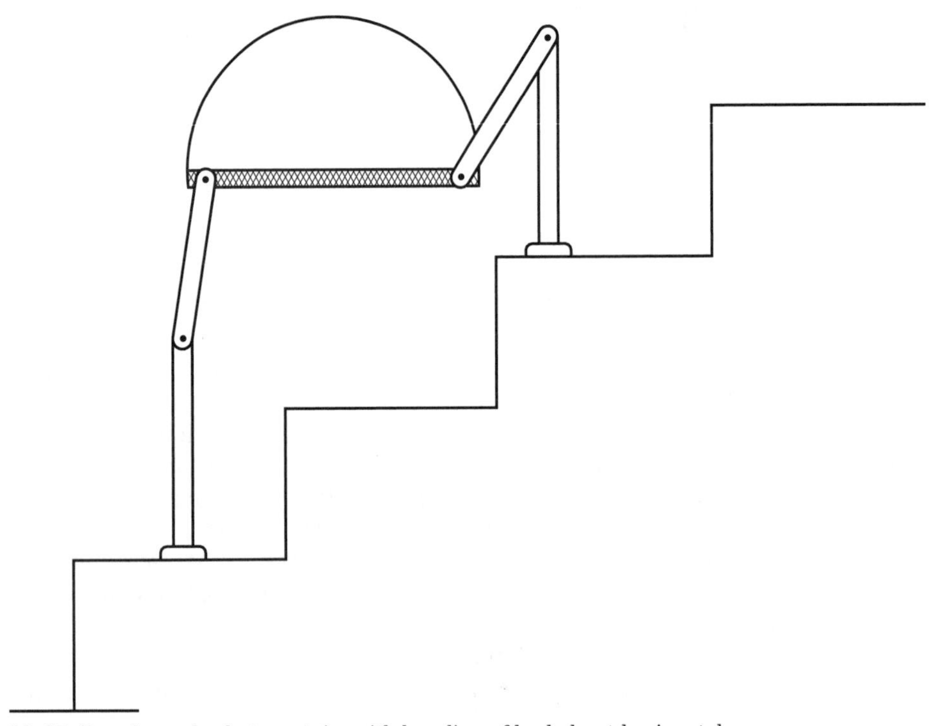

13-15 Four-legged robot on stairs with baseline of body kept horizontal.

the foot pads make good contact. If each leg has three segments, with the short lowest segment existing only to ensure good alignment of the foot pad, we will significantly increase the number of options that we can explore to find a sequence that negotiates the stairs while maintaining static stability throughout the process.

The robot shown measures approximately 13 inches from front to back, the upper leg segments are about 9.4 inches long, and the lower leg segments are about 11 inches long. The stair problem becomes very different for robots which are smaller or larger than this. If the robot is too small to reach from one tread to the next, the stair-climbing problem becomes a wall-climbing problem.

In both Fig. 13-14 and Fig. 13-15, the robot is depicted as spanning three steps: one step supporting the rear legs, one step under the body, and one step supporting the front legs. If we think about it for a few minutes, it becomes obvious that the robot will never make any progress as long as it is constrained to span exactly three steps. It must have at least one of the following capabilities:

- Enough "reach" to advance to front legs by one step, thus temporarily spanning four steps before advancing the rear legs and returning to a span of three steps.
- Enough agility to advance the rear legs, thus temporarily spanning two steps before advancing the front legs and returning to a span of three steps. This agility to advance the rear legs includes the unstated requirement that wherever the rear legs get placed, the center of gravity remains in a stable position relative to the points of ground contact.

Let's put three-segment legs on the level-platform robot from Fig. 13-15, and try this second approach to stair climbing. The modified robot with more specific dimensioning is shown in Fig. 13-16. One possible sequence for climbing up by one step goes as follows:

1. The starting position is shown in Fig. 13-17.
2. Lift the right rear leg from step 1 and place it on step 2 as shown in Fig. 13-18.
3. Lift the left rear leg from step 1 and simultaneously extend the remaining legs to the configuration shown in Fig. 13-19. Place the left rear leg on step 2 as shown.
4. Lift the right front leg from step 3 and place it on step 4 as shown in Fig. 13-20.
5. Lift the left front leg from step 3 and place it on step 4 as shown in Fig. 13-21.
6. At this point it would be possible to slide the right rear leg forward on step 2 and flex all legs to return to the configuration of Fig. 13-17 (only one step higher). However, it would be more efficient to lift the right rear leg from step 2, and then place it on step 3 while simultaneously flexing the other legs to obtain the configuration shown in Fig. 13-18.

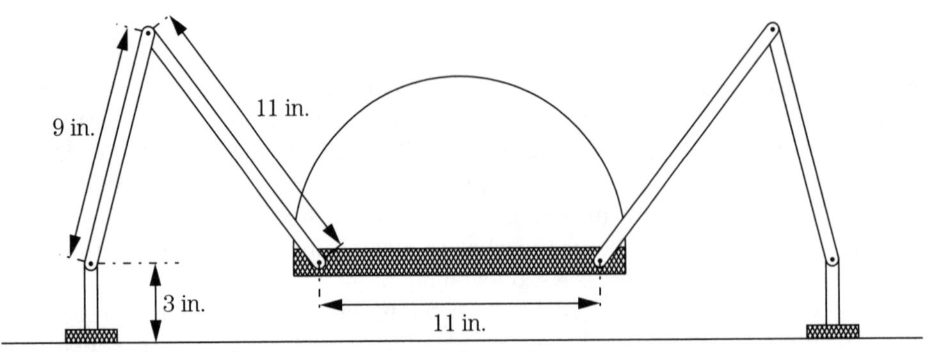

13-16 Hypothetical robot for stair-climbing analyses.

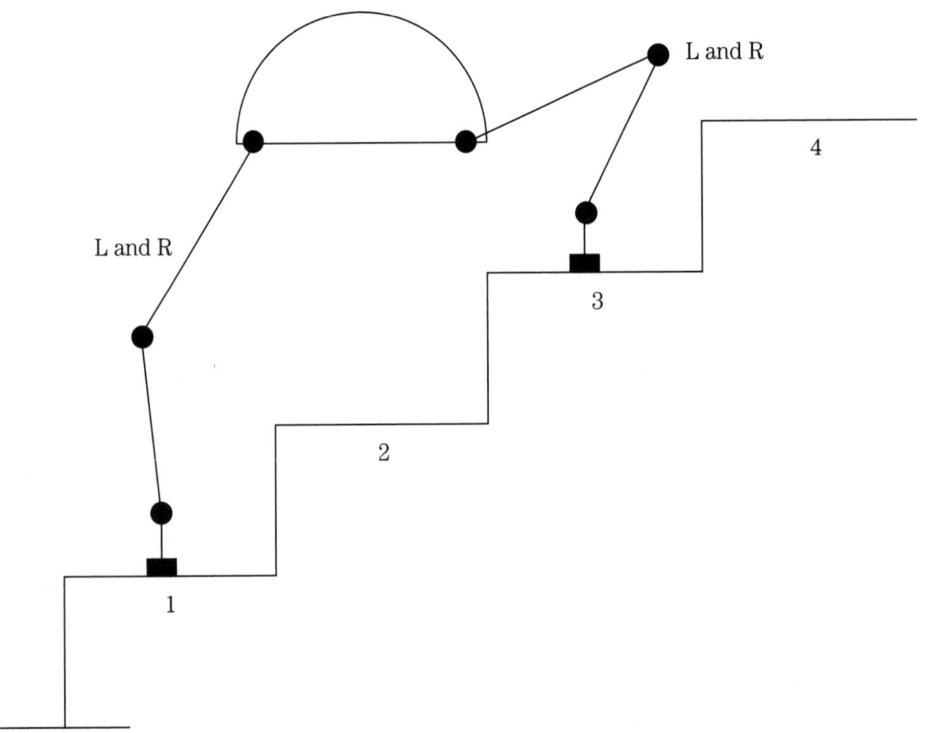

13-17 Starting position for stair-climbing sequence.

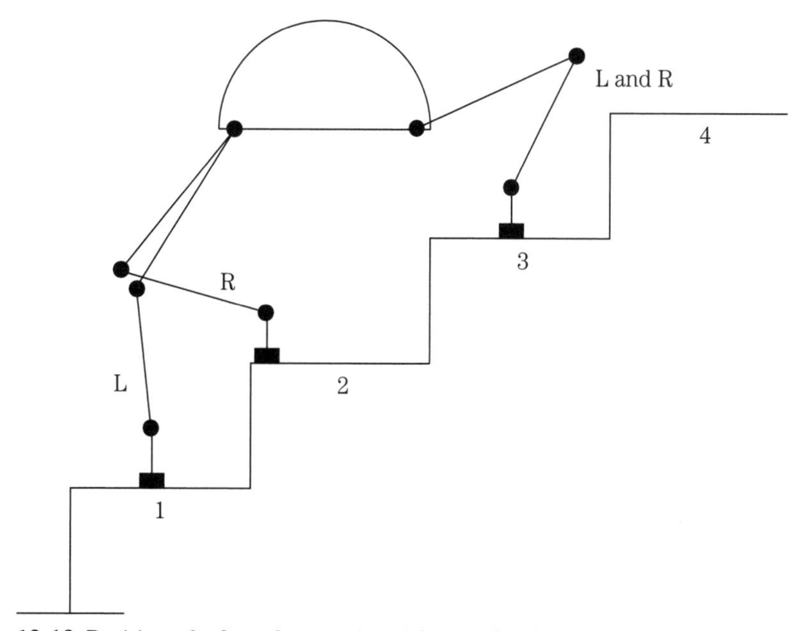

13-18 Position of robot after moving right rear leg from step 1 to step 2.

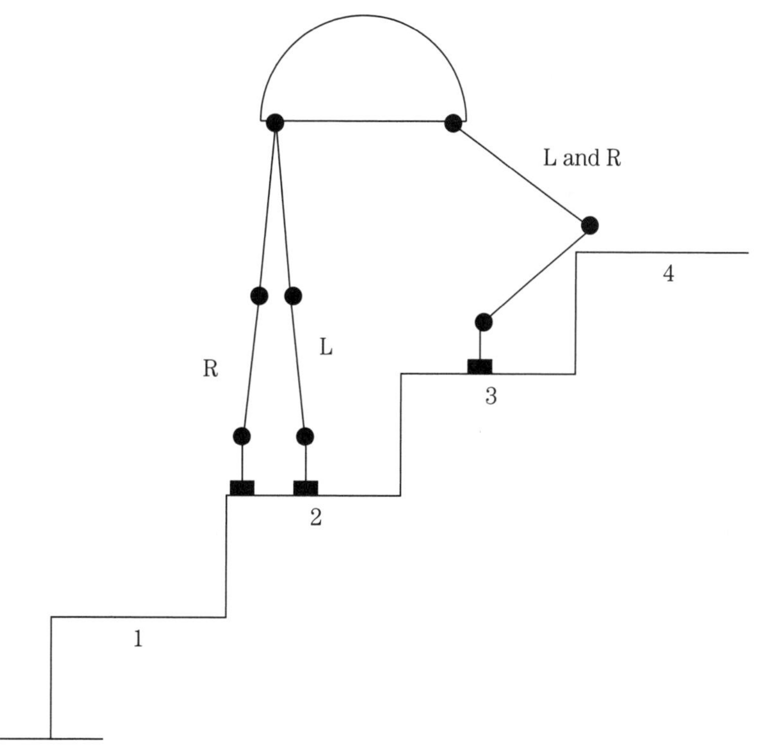

13-19 Position of robot after moving left rear leg from step 1 to step 2 and unflexing all legs.

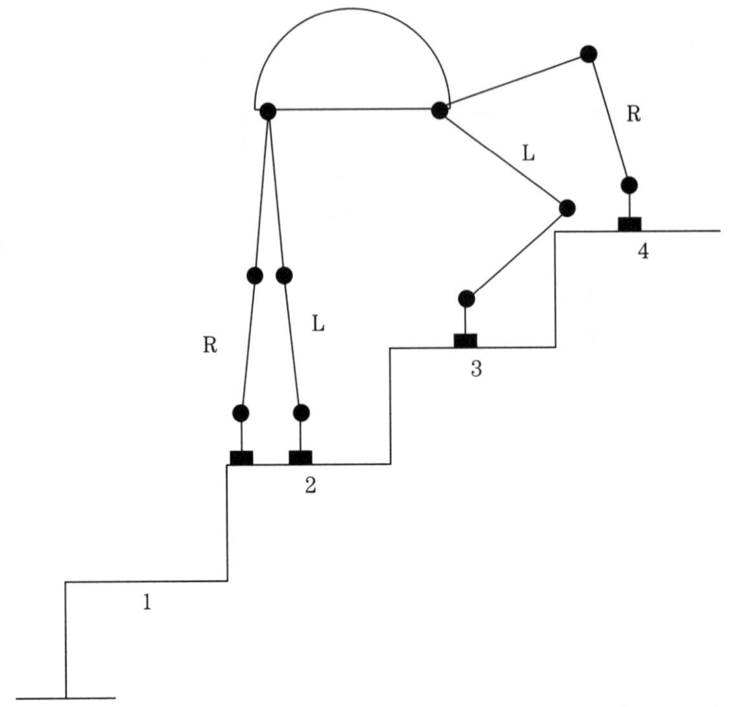

13-20 Position of robot after moving right front leg from step 3 to step 4.

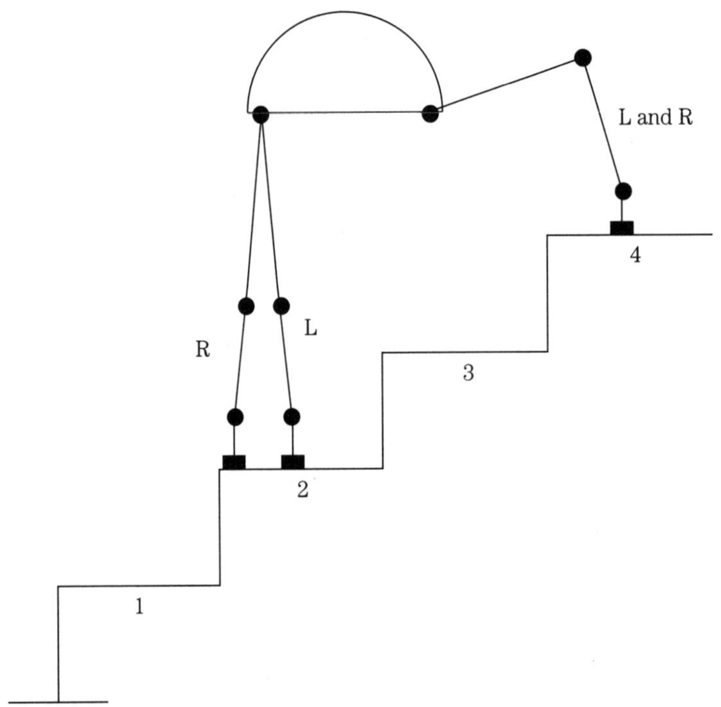

13-21 Position of robot after moving left front leg from step 3 to step 4.

Index

D

E

F

G

H

I

R

S

T

U

V

W

Y

About the author

C. Britton Rorabaugh is a working engineer who holds BSEE and MSEE degrees from Drexel University. His previous books include *Digital Filter Designer's Handbook*, *Circuit Design and Analysis*, and *Communications Formulas and Algorithms*, published by McGraw-Hill, and *Data Communications and LAN Handbook* and *Signal Processing Design Techniques*, published by TAB Professional and Reference Books.

Other Bestsellers of Related Interest

The McGraw-Hill Illustrated Encyclopedia of Robotics and Artificial Intelligence
—Stan Gibilisco
With its 500 alphabetically arranged, densely-illustrated articles and extensive cross-referencing, this is the first encyclopedia devoted solely to robotics and artificial intelligence. Some topics include: robots and automation, robots and AI in medicine, military use of robots, robotics and security, robotic space probes, and more. Gibiliso also includes a comprehensive bibliography to help readers locate additional information.
0-07-023614-3 $24.95 Paper
0-07-023613-5 $40.00 Hard

Home Remote Control and Automation Projects, 2nd Edition
—Delton T. Horn
Fifteen all-new projects expand this outstanding collection that made the first edition a worldwide favorite among hobbyists. You'll find a complete selection of how-to's for making: door and window controllers, liquid monitors and controllers, stereo and TV projects, telephone-related projects, switching units, and more. For all projects, Horn provides you with complete instructions, wiring diagrams, and illustrations for building.
0-07-157726-2 $21.95 Paper

Build Your Own Intelligent, Mobile, SPACE Robot
—Steven James Montgomery
Recent advances in mobile robotics technology have helped make it one of the most popular and exciting fields in the world of electronics. This groundbreaking guide takes readers step-by-step through the entire process of designing and constructing a SPACE (self-programming, autonomous, computer-controlled, evolutionary/adaptive) mobile robot. Also included are a list of robot parts suppliers and a reference bibliography arranged by subject.
0-07-042946-4 $29.95 Paper
0-07-042945-6 $49.95 Hard

Lasers, Ray Guns, and Light Cannons! Projects from the Wizard's Workbench
—Gordon McComb
For electronics professionals, buffs, students, and the experimenter in everyone, this unique book details approximately 100 interesting build-your-own laser projects, from ray guns and laser light shows to night scopes and laser pointers. Readers will learn how to build laser pistols, laser-based "snooping" systems, a novel laser cannon and Gatling gun, laser tachometer, perimeter burglar alarm, and many other intriguing devices.
0-07-045035-8 $21.95 Paper
0-07-045034-X $36.95 Hard

Analytical Robotics and Mechatronics

—Wolfram Stadler

Written as an instruction to robotics and mechatronics, and built on a background of dynamics, circuits, and system analysis, it unifies these separate components and offers a just-in-time approach with any additional material—covering the material when and where it is needed.

0-07-060608-0 $53.44 Hard

Step Into Virtual Reality

—John Iovine

Step Into Virtual Reality is the first book to teach readers how to make their own VR equipment and interfaces to create and enter a VR environment. Projects include a Head-Mounted Display (HMD), LCD stereoscopic glasses, data glove, robot car explorer, tactile sensors, and a VR exercise bike. One of the book's most extraordinary features is included on the accompanying disk. By using the disk and an HMD unit, homebound exercise cyclists can transport themselves to a cyberspace version of the Grand Canyon.

0-07-911906-9 $32.95 Paper (w/disk)
0-07-911905-0 $49.95 Hard (w/disk)